VII International Conference on Solar-Terrestrial Relations and Physics of Earthquake Precursors 2016

E3S Web of Conferences Volume 11 (2016)

Paratunka, Russia
29 August - 2 September 2016

Editor:

B.M. Shevtsov

ISBN: 978-1-5108-3239-8

Printed from e-media with permission by:

Curran Associates, Inc.
57 Morehouse Lane
Red Hook, NY 12571

CURRAN ASSOCIATES INC. proceedings.com

Some format issues inherent in the e-media version may also appear in this print version.

Printed by Curran Associates, Inc. (2016)

For additional information, please contact EDP Sciences – Web of Conferences at the address below.

EDP Sciences – Web of Conferences
17, Avenue du Hoggar
Parc d'Activité de Courtabœuf
BP 112
F-91944 Les Ulis Cedex A
France

Phone: +33 (0) 1 69 18 75 75
Fax: +33 (0) 1 69 28 84 91

contact-edps@webofconferences.org

Additional copies of this publication are available from:

Curran Associates, Inc.
57 Morehouse Lane
Red Hook, NY 12571 USA
Phone: 845-758-0400
Fax: 845-758-2633
Email: curran@proceedings.com
Web: www.proceedings.com

TABLE OF CONTENTS

 DOI: 10.1051/e3sconf/20161100001

Russian Academy of Sciences
Far Eastern Branch
Institute of Cosmophysical Research and Radio Wave Propagation

SOLAR-TERRESTRIAL RELATIONS AND PHYSICS OF EARTHQUAKES PRECURSORS

VII INTERNATIONAL CONFERENCE

v. Paratunka Kamchatsky kray
29 august – 2 September 2016
REPORTS COLLECTION

Petropavlovsk-Kamchatsky
2016

Preface

Reports contain the results of original research of physical processes in the different layers of the atmosphere, near space and the solid Earth. The questions of influence of solar activity on the magnetosphere and ionosphere of the Earth are discussed. Modern methods of measurement, processing and exchange of geophysical data are represented.

Key words: physics of atmosphere, magnetosphere, geophysical fields, earthquake precursors.

Conference was hold under Financial Support of RFBR (№ 16-05-20521 Г)

DOI: 10.1051/e3sconf/20161100001

Scientific committee

edp sciences

E3S Web of Conferences

Statement of Peer review

In submitting conference proceedings to E3S Web of Conferences, I certify to the Publisher that I adhere to the **Policy on Publishing Integrity** of the journal in order to safeguard good scientific practice in publishing.

1. All articles have been subjected to peer review administered by the proceedings editors.
2. Reviews have been conducted by expert referees, who have been requested to provide unbiased and constructive comments aimed, whenever possible, at improving the work.
3. Proceedings editors have taken all reasonable steps to ensure the quality of the materials they publish and their decision to accept or reject a paper for publication has been based only on the merits of the work and the relevance to the journal.

Title, date and place of the conference

VII INTERNATIONAL CONFERENCE
"SOLAR-TERRESTRIAL RELATIONS AND PHYSICS OF EARTHQUAKES PRECURSORS"
29 august - 2 september 2016
Kamchatka region, Elizovskiy district, Paratunka Mirnaya str., 7
Institute of Cosmophysical Research and Radio Wave Propagation FEB RAS

Proceedings editor(s):

Professor B.M. Shevtsov,
Institute of Cosmophysical Research and Radio Wave Propagation FEB RAS

Date and editor's signature

06.10.2016

E3S Web of Conferences **11**, 00002 (2016) DOI: 10.1051/e3sconf/20161100002
Solar-Terrestrial Relations and Physics of Earthquakes Precursors

Calculation of seismic regime parameters on the basis of a probabilistic model of Kamchatka earthquake catalogue

Vadim Bogdanov[1,*] and *Aleksey* Pavlov[1,**]

[1] *Institute of Cosmophysical Research and Radio Wave Propagation FEB RAS, Paratunka, Russia*

Abstract. In the present paper on the basis of a probabilistic interpretation of the earthquake repetition law, simple formulas were obtained which allow us from the known number of earthquake events in a representative interval of energy class (magnitude) to estimate the number of seismic events falling within an unrepresentative interval and the expected time of an earthquake with a magnitude exceeding the maximum registered values for the period of instrumental observations.

1 Introduction

The character of Kamchatka seismicity is determined by interaction of several lithospheric plates: Pacific, Okhotsk, Eurasian, North-American and Bering. The majority of earthquakes and the strongest events are associated with Pacific plate subduction. This process is accompanied by tectonic stress accumulation and relaxation and manifests in a large number of earthquakes, the hypocenters of which are associated with the descending oceanic plate. Earthquakes occurring in Kamchatka have different tectonic nature [1]. Figure 1 illustrates six tectonic-geographical zones (structures) of Kamchatka region which have significantly different characteristics of seismicity. S_1, S_2, S_3 and S_4 zones have complicated three-dimensional configuration and sloping edges in depth. S_5 and S_6 zones do not contain deep earthquakes and have vertical edges.

On the basis of a probabilistic approach proposed in [2, 3], it is possible to analyze the seismic regime for tectonic-geographical zones in Fig. 1. On the basis of probabilistic interpretation of earthquake repetition law graph suggested in [2] and considered in detail in [3, 4], it is possible to obtain simple relations allowing:

- from the known number of earthquakes in a representative energy class (magnitude) to calculate the number of events falling within a unrepresentative interval;
- on the considered area to estimate the expected time of an earthquake with magnitude exceeding maximum registered values for the instrumentation period of observations.

Earthquake repetition law in a probabilistic representation for energy class K distribution density of earthquake events has the following form [2]

$$f(K^{1,2}) = f(K_0^{1,2}) \cdot 10^{-\gamma_{1,2}(K^{1,2}-K_0^{1,2})} \tag{1}$$

*e-mail: vbogd@ikir.ru
**e-mail: pavlov@ikir.ru

DOI: 10.1051/e3sconf/20161100002

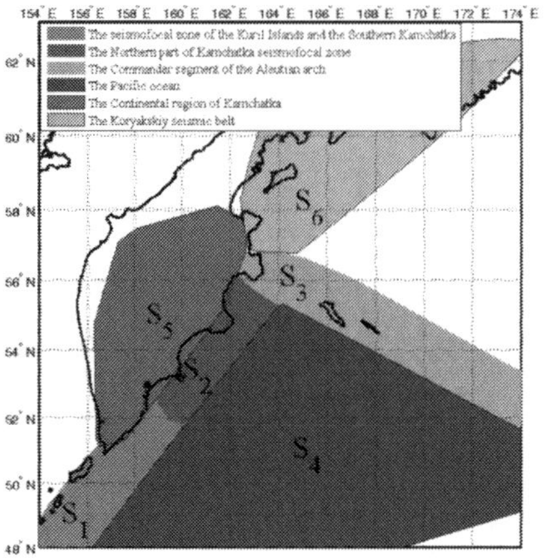

Figure 1. Surface-section of seismicity zones for Kamchatka region

or in a double logarithmic scale the dependence (1) may be presented as

$$\lg f(K^{1,2}) = \lg f(K_0^{1,2}) - \gamma_{1,2}(K^{1,2} - K_0^{1,2}) \tag{2}$$

Expressions (1) and (2) are written for the following ranges of changing of K: $9.5 \leq K_1 < 12.5$ with γ_1 ($K_0^1 = 9.5$) and $12.5 \leq K_2 < 15.5$ with γ_2 ($K_0^2 = 12.5$) and $f(K_0)$ denotes initial distribution density calculated in the middle of the interval $K_0 \pm \Delta K/2$.

2 Calculation of unrepresentative energy class event number

It was shown by the calculations on the example of Kamchatka earthquake catalogue [3], that it does not depend on the fact if a catalogue is cleared from aftershocks or not, for the fixed period of time T and a given initial interval $K_0 \pm \Delta K/2$, the distribution densities $f(K)$, determined by (1), almost do not depend on the considered area where epicenters of events are located. Spread of values appears only in the result of different number of events falling within the considered area. The larger is the number of events, the more stable is the distribution $f(K)$. It is achieved either by the increase of an area for a given time period T or by the increase of a period itself for a fixed area. Table 1 shows the probability values for the earthquakes to fall within the interval of the energy class K, calculated for the period T from 01.01.1962 to 01.01.2008 and averaged for $S_1, S_2, \ldots, S_6$ zones for the uncleared catalogue of Kamchatka earthquakes [5].

For different initial K_0 the following values $P(K_0)$ were obtained when $\Delta K = 1$ (Table 1): $P(K_0 = 9.5 \pm 0.5) = 0.658$, $P(K_0 = 10.5 \pm 0.5) = 0.692$, $P(K_0 = 11.5 \pm 0.5) = 0.713$, $P(K_0 = 12.5 \pm 0.5) = 0.716$. The average value of the probability $P_{av}(K_0)$ for different initial K_0 is equal to 0.6947.

Figure 2 shows the dependences (2) for different K_0, equal to 9.5, 10.5, 11.5 and 12.5, calculated for the tectonic-geographic S_2 zone. On the basis of the dependence (2) the values of incline angles γ_1 and γ_2 of the earthquake repetition law were calculated (Table 2) for two ranges of the energy class $9 \leq K < 13$ and $12 \leq K < 16$, correspondingly. The values of γ (Table 2) presented in the paper [1] and the values of γ_1 calculated on the basis of probability distributions almost coincide.

Comparing different values of probabilities showed in Table 1, one may note that with the increase of initial K_0 the values of the obtained probabilities have the tendency of gradual increase. It is

Table 1. Averaged distributions of probabilities for different initial values K_0

Intervals of energy class K_i	Average $P(K_i)$ for the areas of $S_1, S_2, \ldots, S_6$, $K \geq 9$	Average $P(K_i)$ for the areas of $S_1, S_2, \ldots, S_6$, $K \geq 10$	Average $P(K_i)$ for the areas of $S_1, S_2, \ldots, S_6$, $K \geq 11$	Average $P(K_i)$ for the areas of $S_1, S_2, \ldots, S_6$, $K \geq 12$
9.5 ± 0.5	0.658			
10.5 ± 0.5	0.234	0.692		
11.5 ± 0.5	0.076	0.218	0.713	
12.5 ± 0.5	0.023	0.064	0.205	0.716
13.5 ± 0.5	0.007	0.019	0.059	0.208
14.5 ± 0.5	0.0017	0.005	0.017	0.0602
15.5 ± 0.5	0.00072	0.002	0.00523	0.016

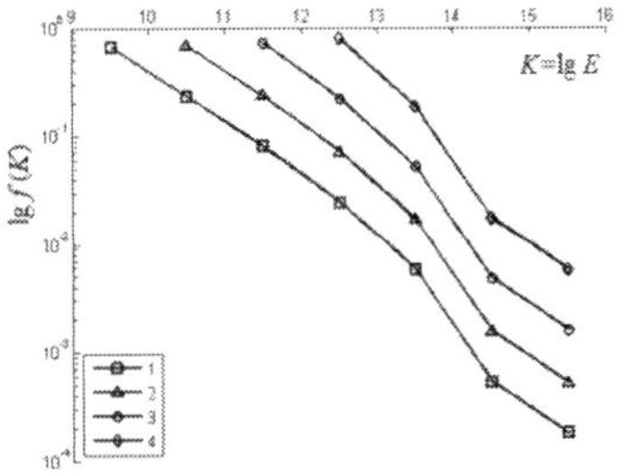

Figure 2. $f(K)$ dependence in a double logarithmic scale (relation (2)) for S_2 zone for different initial values K_0: 1) $K_0 = 9.5 \pm 0.5$, 2) $K_0 = 10.5 \pm 0.5$, 3) $K_0 = 11.5 \pm 0.5$, 4) $K_0 = 12.5 \pm 0.5$.

Table 2. Inclination angle values of the graph of earthquake repetition law

	S_1	S_2	S_3	S_4	S_5	S_6
$\gamma, 9 \leq K < 13$	-0.5	-0.43	-0.45	-0.7	-0.53	–
$\gamma_1, 9 \leq K < 13$	-0.476	-0.425	-0.456	-0.623	-0.589	-0.434
$\gamma_2, 12 \leq K < 16$	-0.745	-0.726	-0.595	–	–	-0.309

explained by the fact that with the decrease of K_0, some difficulties on registration of events appear. They are determined by unbranched network of seismic stations, external disturbing factors and so on, which eventually result in the gaps of occurred earthquakes. Thus, the graph fall of the repetition law for small values of K, and, correspondingly, the probability decrease occur due to the decrease of the registered number of events falling within the unrepresentative region of energy class values. On the other side, with the increase of initial value of the complete group of real events decreases and the part of events in the initial chosen interval increases and the probability in frequency presentation grows.

Calculation of distribution density on the basis of real registered events with energy less than a representative one and its comparison with $f(K_i)$, calculated on the basis of the relation (1), give disagreement. Thus, the task appears to obtain a formula for calculation of event number in energy class unrepresentative interval on the basis of known event number and, consequently, distribution density in representative class. Hence, if for period T and for area S within the energy class representative interval $K_0 \pm 0.5$, $n_0(\Delta K)$ earthquakes were registered when their total number is N_Σ, forming an event complete group with energy classes from $K_0 \pm \Delta K$ to $K_{max} \pm \Delta K$, than the number of events n_{-1}, occurred in the next unrepresentative interval with $K_{-1} < K_0$, may be determined by the relation (3).

$$n_{-1} = \frac{N_\Sigma \cdot P_0(K_0 \pm \Delta K/2)}{1 - P_0(K_0 \pm \Delta K/2)} \tag{3}$$

It is assumed that with K value decrease in unrepresentative class region,γ_1 does not change. Now we determine the number of earthquakes which occurred in the northern part of Kamchatka seismofocal zone (S_2 zone) for the period T (from 01.01.1962 to 01.01.2008) and falling within the unrepresentative interval of the energy class 8.5 ± 0.5. Since the total number of events with $K \geq 9$, which occurred in S_2 zone for the period T, is $N_\Sigma = 11323$, and the probability to fall within the initial interval is $P(9.5 \pm 0.5) = 0.6582$, then according to the recurrent relation (3) we obtain $n_{-1} = 21805$. At the same time according to the earthquake catalogue, the total number of registered events for $T = 46$ years on the area S_2 and falling within the interval 8.5 ± 0.5 equals 16893. Similarly, on the basis of (3) the number of events was calculated that fall into energy class interval 8.5 ± 0.5 for other considered zones S_i. The calculation results are shown in Table 3.

Table 3. The number of events in an unrepresentative range of energy class

Zone	S_1	S_2	S_3	S_4	S_5	S_6
$n(8.5 \pm 0.5)$, (catalogue)	7308	11323	3177	1534	2348	710
$n(8.5 \pm 0.5)$, (relation (3))	10566	21805	6361	3912	3698	1000

3 Calculation of the expectance period of a strong earthquake

The following expression was obtained in the paper [2] on the basis of relation (1)

$$\lg N = A - bM \tag{4}$$

where $A = \lg N(M_0) + bM_0 = \lg[10^{bM_0} \cdot N(M_0)]$ and $b = 1.5\gamma$, $N(M_0)$ is the number of events falling within the interval $M_0 \pm \Delta M/2$, $M_0 = (K_0 - 4.6)/1.5$ ($M_0^1 = 3.27$ for $K_0 = 9.5$, $M_0^2 = 5.27$ for $K_0 = 12.5$, $\Delta M = 0.67$ for $\Delta K = 1$). The relation between K class and M magnitude for Kuril-Kamchatka earthquakes was specified by $K = 4.6 + 1.5M$. When there is no enough statistics in the region of high M values, as $\lg N$ becomes negative, and the time of expectance exceeds the period of instrumentation observations, it is possible to estimate this time by relation (4) for magnitude higher values compared to already registered ones for the period of instrumentation observations. We suppose that spatial and energy characteristics of Kamchatka earthquakes for 01.01.1962–01.04.2016 at the area reflect, in average, its seismic regime. Thus, the average number of events, the epicentres of which are located on area S_7 (Fig. 3), calculated for a long period of observations (several decades, for example), describes quite well the average regime for a longer period. In this case, from the equality of average calculated for the periods of instrumentation observations T_{ins} and the required period T_{req} during which one event may occur with maximum magnitude M_{max}, we obtain

$$T_{req} = \frac{10^{1.5\gamma(M_{max} - M_0^2)}}{N(M_0^2)} T_{ins} \tag{5}$$

According to (5) the expected time for Kamchatka region for three magnitudes $M_{max}^1 = 8.5$, $M_{max}^2 = 9.0$ and $M_{max}^3 = 9.5$ is estimated. The results of data processing from the catalogue for

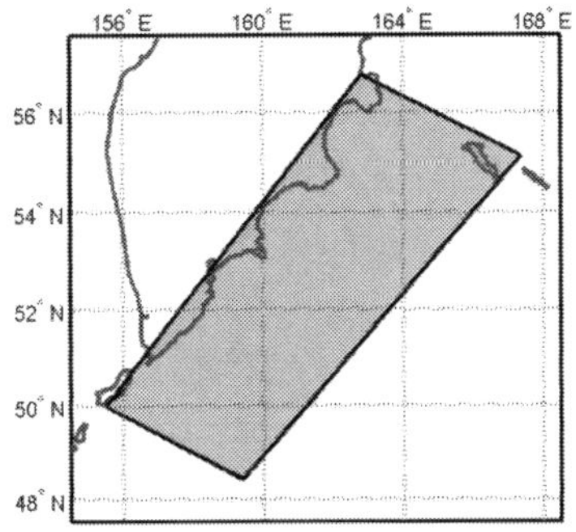

Figure 3. S_7 zone

Kamchatka region for zone S_7 for the period of T_{ins} = 54 years (from 01.01.1962 to 01.04.2016): N_Σ = 27117 is the summary number of events of the catalogue with the energy class within the interval from 9.5 ± 0.5 to 15.5 ± 0.5 (for the magnitudes from 3.27 ± 0.335 to 7.27 ± 0.335); $P(5.27 \pm 0.335) = 0.024$ is the probability for seismic events to fall within the interval $M_0^2 \pm 0.335$; $N(M_0^2) = P(5.27 \pm 0.335) \times N_\Sigma = 0.024 \times 27117 = 662$ is the number of events registered on S_7 area for T_{ins} within the magnitude interval $M_0^2 \pm 0.335$; $|\gamma| = 0.64$ is the incline coefficient calculated by the least square method. Calculation on the basis of the given data results in the following values of the expected periods with corresponding magnitudes of events: for $M_{max}^1 = 8.5$ expected period $T_{req}^{8.5} \approx 100$ years; for $M_{max}^2 = 9.0$ expected period $T_{req}^{9.0} \approx 300$ years; for $M_{max}^3 = 9.5$ expected period $T_{req}^{9.5} \approx 960$ years.

4 Conclusion

It was shown that for the considered zones of $S_1, S_2, \ldots, S_6$ and different values of the initial interval of the energy class $K_0 \pm 0.5$, for K_0 = 9.5, 10.5, 11.5 and 12.5, the probability values $P(K_0 \pm 0.5)$ differ little from each other and tend to the average value 0.6947.

On the basis of recurrence relation (3) the estimation of events which must occur in the unpresentative energy class $K_0 \pm 0.5$ was calculated for the zones of $S_1, S_2, \ldots, S_6$.

On the basis of relation (5) the estimation of time of a strong earthquake for zone S_7 was calculated. The periods of expectance of seismic events with magnitudes 8.5, 9.0 and 9.5 were 100, 300 and 960 years, respectively.

References

[1] V.I. Levina, A.V. Lander, S.V. Miryushkina, A.Yu. Chebrova, Journal of Volcanology and Seismology **7**, 37-57 (2013)

[2] V.V. Bogdanov, Doklady Akademii Nauk **408**, 393–397 (2006)

[3] V.V. Bogdanov, A.V. Pavlov, A.L. Polyukhova, Journal of Volcanology and Seismology **4**, 412-422 (2010)

[4] V.V. Bogdanov, A.V. Pavlov, A.L. Polyukhova, Izvestiya vysshikh uchebnykh zavedeniy. Severo-Kavkazskiy region. Seriya: Estestvennye nauki **1**, 44-48 (2012)

[5] http://www.emsd.ru/sdis/earthquake/catalogue/catalogue.php

[6] S.A. Fedotov, *Energy classification of Kuril-Kamchatka earthquakes and magnitude problem* (Nauka, Moscow, 1972)

Impact of cyclones over kamchatka on electron distribution in the ionosphere

Vadim Bogdanov[1,⋆], *Alexandr* Kaysin[2,⋆⋆], and *Alexandr* Romanov[3,⋆⋆⋆]

[1] *Institute of Cosmophysical Research and Radio Wave Propagation FEB RAS, 684034, Paratunka, Mirnaya str., 7, Russia*
[2] *Institute of Cosmophysical Research and Radio Wave Propagation FEB RAS, 684034, Paratunka, Mirnaya str., 7, Russia*
[3] *OAO "Russian Space Systems"*

Abstract. The paper presents the results of investigation of cyclone impact on ionosphere parameters. Ionosphere state was observed by automatic sounding equipment applying low-orbital navigation spacecrafts in the conditions of low seismicity.

1 Introduction

Kamchatka region is located in the area of junction of the Asian continent and the Pacific Ocean, and it is characterized by increased seismic activity. The whole territory is within the zone of intensity 5-9 earthquakes (12-grade seismic intensity scale). Thus, the search for the precursors of catastrophic seismic events to make their forecast is very important. The well-known factor is the relation of the processes occurring on the Sun and in the Earth's lithosphere with the processes in the ionosphere manifesting as registered precursors (see, for example, the monograph [1] and the references there). However, registration of a reliable forecast feature of a seismic event in the ionosphere parameter behavior faces some difficulties which are associated with the fact that the Earth's atmosphere is a dynamic system affected by different physical external processes. Consequently, ionospheric parameter variations cannot be explained completely not taking into account the influence of the atmosphere lower layers. Thus, physical processes generated in the lower and middle atmosphere may affect the parameters in the upper layers including ionosphere. Such atmospheric phenomena as cyclones may have some impact on the ionosphere state in a varying degree. Since the Kamchatka territory is washed by the Pacific Ocean, by the Okhotsk and the Bering Seas, which are the zones of increased cyclone activity, Kamchatka may be considered as a field test site for investigation of the impact of cyclone activity on the processes in the ionosphere when the corresponding measurement systems are available. By the present moment, a cycle of work has been carried out. The relations between wave disturbances in the atmosphere and weather conditions were under investigation. Wave disturbances in ionospheric parameters during storm passage were observed [2]. Quasiperiodic vibrations were

⋆e-mail: vbogd@ikir.ru
⋆⋆e-mail: kaisin@mail.ru
⋆⋆⋆e-mail: romulas@mail.ru

detected during the propagation of a strong thunderstorm [3]. At present time, there are papers devoted to the investigation of possible impact of tropical cyclones on different layers of the ionosphere. Such disturbances were registered by different methods. Mid-scale traveling ionospheric disturbances (TID) with the period of about 20 minutes were detected during propagation of storms by the measurements of Doppler frequency shift in the ionosphere [4]. It was shown that at the distance of about 3-5 thousand of kilometers from a cyclone center, changes of f_0F2 parameter by 10-20% were observed by ionosondes of vertical and oblique sounding [5]. The authors of the papers think that increase [6] or decrease [7] of the critical frequency f_0F2 may be associated with a delay of the moment of measurement relatively the time of cyclogenesis. In the papers [8, 9], time intervals of increase of signal spectrum power at definite frequencies were detected. Increase of acoustic gravitational wave (AGW) activity by 20-30% was determined at the stage of fast change of hurricane intensity by GPS measurements. However, there are some publications where ionosphere response was not detected [11]. In this paper, a cycle of investigations of tropospheric cyclone effect on ionopheric parameters is suggested on the example of winter and spring cyclones. This survey applies the data obtained by the method of tomographic sounding by a chain of tomographic stations located on Kamchatka peninsula. The peculiarity of this work is the location of the receiving stations for tomographic sounding in the area of cyclone effect.

2 Description of the method

Radio tomographic sounding of the ionosphere is carried out over the territory of Kamchatka peninsula by ray radio tomography. The tasks of ray radio tomography are practically solved by satellites and several receivers. Several stations for registration of navigation signals are arranged on the Earth surface in the plane which corresponds to the angles of orbital inclination of satellites. In case of application of "Tsikada/Tranzit" space systems, the orbital inclination of which is $83° - 90°$ degrees, ground receiving stations for navigation signals should be arranged in meridian direction. On the ground, electron content on the propagation path of a navigation signal may be estimated by the phases of signals with the frequencies of 150 and 400 MHz. During the time of satellite flying over the extended system of ground stations, a series of measurements is carried out. For each station of registration of a navigation signal we determine

$$TEC = \int_p N_e ds \qquad (1)$$

, (where TEC is the total electron content along the path of signal propagation between a source and a receiver p, N_e is the electron content), on an element of propagation path ds. Then the tomographic crosscut is reconstructed by an inverse problem [12, 13]. Fig. 1 shows a scheme of realization of the method of ionosphere tomographic sounding which allows us to restore electron concentration in the upper atmosphere. Low-orbital navigation satellites of "Kosmos" type were used as sources of coherent signals. It allows us to reproduce the concentration vertical distribution along the track of a satellite with good resolution (25 km height and 50 km latitude). Ground receivers of signals are located in Paratunka (*Par*), Mil'kovo (*Ml*), Esso (*Ess*) in sub-meridian direction (158°31′ E). The receiving stations operate automatically.

3 Analysis of experimental results

To study the ionosphere response, we chose the cyclones which passed over Kamchatka within the period from January 2012 to February 2014 (Table 1) in the conditions of calm and moderately calm

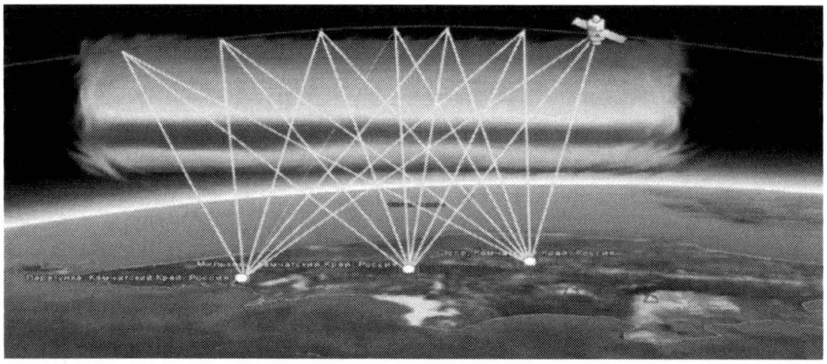

Figure 1. Scheme of arrangement of receiving stations and sounding signal propagation.

magnetic state (K ≤ 17). During the passage of the cyclones there were no strong earthquakes in Kamchatka region and the seismic regime corresponded to the background mode. It gives us grounds to hope that if there are effects of cyclone influence on the ionosphere, this influence is not masked by the processes occurring in the lithosphere of the seismically active region. The left part of the Table 1 defined as "Day with cyclone" shows the date of its passage, time of satellite fly-over (UT), total index of magnetic activity (K) on the mentioned date at "Paratunka" magnetic station. The second part of the Table ("Day without cyclone") gives the information on the closest to the cyclone reference day for which electron meridian distribution was determined for comparison. The time of satellite fly-over on the reference day was chosen to be the most close to the time of fly-over on the day of a cyclone. As an example, Fig. 2 illustrates synoptic charts and tomograms of electron sub-meridian height distributions for two days with the cyclone (January 18, 2013) and without it (January 12, 2013). The tomograms show the latitude in degrees and the stations of satellite data registration *Par*, *Ml*, *Ess*, along the x-axis and the height in km along the y-axis. The scale of electron concentration distribution is to the right from the tomograms. Tomograms are taken within the latitude range of 48.01° – 61.52° N.

Table 1.

Day with cyclone			Day without cyclone		
Dat	Time of satellite fly-over, UT	Total K-index	Date	Time of satellite fly-over, UT	Total K-index
13.01.2012	15:10	9	09.01.2012	15:06	10
01.04.2012	12:49	14	29.03.2012	12:32	4
18.01.2013	16:00	16	12.01.2013	16:20	2
03.02.2013	18:09	7	31.01.2013	18:32	1
17.02.2013	12:33	17	12.02.2013	13:07	9
18.02.2013	09:20	5	15.02.2013	09:46	7
31.01.2014	14:15	0	25.01.2014	14:40	13
05.02.2014	17:05	9	25.01.2014	17:33	13

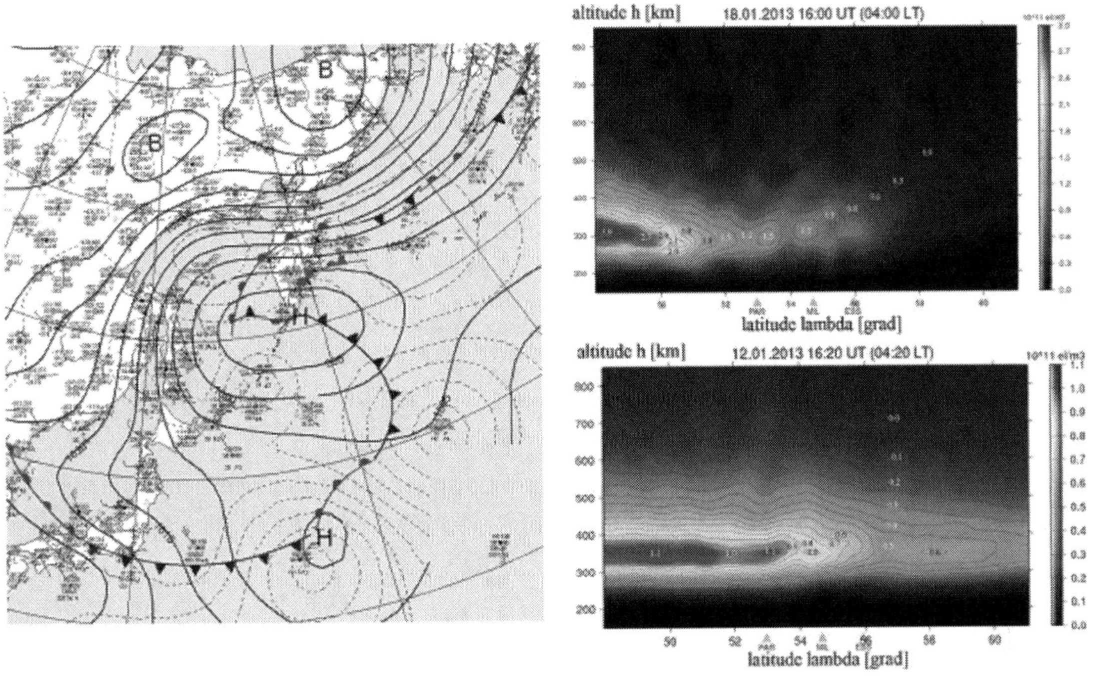

a) b)

Figure 2. Cyclone on January 18, 2013 a) Synoptic chart corresponds to 12 UT. Eastern wind reached 30 m/s at the height of 5 km. Cyclone center was located in the Southern part of the peninsular at ~ 48° N (12 UT) b) On January 18, 2013 at 16:00 UT ionospheric plasma turbulization is observed with clearly defined plasma formation within the latitude range of 48° – 49° N and the height range of 290–350 km with the latitude increase (January 12, 2013 is the reference day).

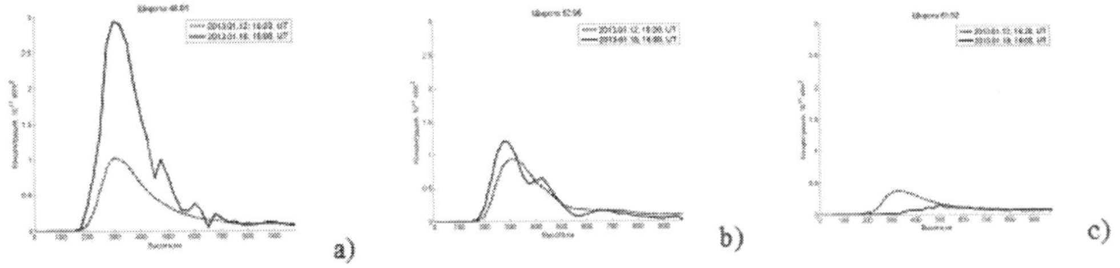

Figure 3. Profile of electron density distribution in the ionosphere on January 18, 2013 at 16:00 UT (04:00 LT) (solid line) for three latitudes in comparison to the reference day on January 12, 2013 (dashed line).

Table 2 shows the highest values of electron concentration for the three chosen latitudes of 48.01° N, 52.96° N and 61.52° N for the cyclone under analysis and the corresponding reference day.

Table 2. Highest value of electron concentration of the ionosphere, $el/m^3(\times 10^{11})$

Satellite fly-over/ Latitude	48.01° N	52.96° N	61.52° N
12.01.2013; 16:20 UT	1,0287	0,9313	1,0287
18.01.2013; 16:00 UT	2,9366	1,2030	0,1486

Conclusions

The analysis of cyclone development showed that:

- Increases in electron concentration are observed for five out of eight cyclones compared to the reference days near the cyclone centers.
- The highest turbulization of the ionosphere is observed just over the cyclone center area which is characterized by steep gradients in electron distributions.
- Electron concentration change in the ionosphere may be explained by mechanical momentum transfer from the lower layers of the ionosphere and neutral atmosphere [14] to the heights of F-layer and by cyclone-induced AGW effect.
- Further investigations of cyclone impact on the ionosphere are necessary to collect statistics, to develop physical models and to determine the mechanism of transfer of cyclone rotational motion momentum to ionospheric heights.

The work is supported by RFBR Grant No. 11-05-00915, by Presidium RAS Program No. 12-1-P22-01 and FEB RAS Program "Satellite monitoring of the Far East for fundamental scientific investigations of FEB RAS".

References

[1] Pulinets S.A. et al., *Prognoz zemletryaseniy vozmozhen?!* (Trovant, 2014) 144

[2] Baker Donald M., Davies Kenneth, Journal of Geophysical Research 70 **Ionospheric Effects Observed around the Time of the Alaskan Earthquake of March 28, 1964**, 2251-2253 (1965)

[3] Prasad S.S., Schneck L.J., Davies K., Journal of Atmospheric and Terrestrial Physics 37 **Ionospheric disturbances by severe tropospheric weather storms**, 1357-1363 (1975)

[4] Xiao Z. et al, Journal of Geophysical Research. Space Physics 112 **Morphological features of ionospheric response to typhoon**, A04304 (2007)

[5] Vanina-Dart L. B., Romanov A. A.,Sharkov E. A, Journal Geomagnetism and Aeronomy. 51 **Influence of a tropical cyclone on the upper ionosphere according to tomography sounding data over Sakhalin Island in November 2007**, 774-782 (2011)

[6] Rice D.D. et al., Journal Radio Science. 47 **Typhoon Melor and ionospheric weather in the Asian sector: A case study** , (2012)

[7] Liu Y.–M., Wang J.–S., Suo Y.–C., Journal Adv. Geosci. 29 **Effects of typhoon on the ionosphere**, 351-360(2006)

[8] Mikhailov Yu.M. et al., Journal Geomagnetism and Aeronomy. 45 **Possible atmospheric effects in the lower ionosphere according to atmospheric radio noise observations on Kamchatka during tropical cyclones**, 778-792(2005)

[9] Chernigovskaya M.A. et al., Journal Issledovanie Zemli iz kosmosa. **Issledovanie svyazi korotkoperiodnykh vremennykh variatsiy parametrov ionosfery v Severo-Vostochnom regione Rossii s proyavleniyami tropicheskikh tsiklonov**, 32-41 (2010)

[10] Zakharov V.I., Kunitsyn V.E, Journal Geomagnetism and Aeronomy. 52 **Regional features of atmospheric manifestations of tropical cyclones according to ground- based GPS network data**, 533-545 (2012)

[11] Afraimovich E.L. et al, Journal Geomagnetism and Aeronomy. 48 **Variations in the total electron content during the powerful typhoon of August 5–11, 2006, near the southeastern coast of China**, 674-679 (2008)

[12] Andreeva E.S., Kunitsyn V.E., Tereshchenko E.D., Journal Geomagnetism and Aeronomy. 32 **Fazoraznostnaya tomografiya ionosfery**, 104-110 (1992)

[13] Romanov A.A. et al, Journal Sovremennye podkhody k sozdaniyu avtomatizirovannoy sistemy registratsii ionosfernykh predvestnikov zemletryaseniy po sputnikovym dannym. **Fazoraznostnaya tomografiya ionosfery**, 167-172 (2006)

[14] Danilov A.D. et al., *Meteorologicheskie effekty v ionosfere* (Gidrometeoizdat, 1987)

Research of variations of ionospheric parameters during the local meteorological phenomena in the Kaliningrad region

Olga Borchevkina[1,*], *Ivan* Karpov[1,2], *Alexandra* Ilminskaya[1], *Yuliya* Kurdyaeva[1], *Alexey* Karpov[1], and *Nina* Korenkova[2]

[1]Immanuel Kant Baltic Federal University, 236006 Kaliningrad, Russia
[2]West department IZMIRAN, 236041 Kaliningrad, Russia

Abstract. The paper presents observations of atmospheric and ionospheric parameters during strong meteorological disturbances (storms) in the Kaliningrad region. The analysis of ionospheric observations has shown that during meteorological storms the amplitude of diurnal variations in TEC decreases to 50 %; and in foF2, to 15 % as compared to quiet days. The revealed changes in ionospheric conditions during meteorological storms are regularly registered and represent a characteristic feature of the meteorological effect on the ionosphere. Modeling studies of the vertical propagation of AGW from the Earth's surface showed that such waves quickly (within ~15 min) reach altitudes of the upper atmosphere (~300 km). The refraction and dissipation of waves in the upper atmosphere produces perturbations of the background state of the atmosphere and gives rise to the waveguide propagation of infrasonic wave components. The observed manifestations of TEC disturbances caused by AGW propagating from the lower atmosphere can be explained by the diurnal variation of the altitude of the ionosphere and the waveguide propagation of infrasonic waves.

1 Introduction

Meteorological processes are an important source of acoustic-gravity waves (AGW) in the atmosphere. Results of multiple experimental and theoretical studies suggest that different dynamic processes in the lower atmosphere and on Earth's surface, which are caused, for example, by meteorological, seismological, and other events, have a significant effect on ionospheric conditions. Assessments of the impact of such processes in the lower atmosphere on ionospheric conditions are reported in [1-4].
In particular, changes in the structure of ionospheric variations during meteorological disturbances are analyzed in [5-9]. Strong meteorological disturbances are characteristic of the climate of the Kaliningrad region. This allows us to identify meteorological effects in changes of ionospheric conditions.

* Corresponding author: opsuslova@gmail.com

2 The model of atmospheric processes

The propagation of AGW in the lower atmosphere is theoretically studied using numerical models that take into account a variety of physical processes involved in the propagation of waves. The analysis of accumulation of commutation errors is studied in details in [10-12]. The constructed method allows the simulation of waves for times up to several tens of hours, and to distances up to several thousand kilometers and from the Earth's surface to the thermosphere heights without significant computational errors. This allows, in particular, computation of the impact of waves on the atmosphere and make simulation of wave propagation with consideration of changing of the medium properties due to transfer of energy of dissipated waves to the medium.

[13] explores methods of mathematical modeling the spread of AGW in the upper atmosphere from the surface source. According to theoretical studies, that above the source at an altitude of ~ 200 km due to the dissipation of the AGW heating a region of the atmosphere with a horizontal scale of about 1000 km with a duration of operation of the power ~ 1 hour. The emergence of the heated region changes the propagation conditions of AGW and leads to the waveguide propagation of waves with periods shorter than the Brunt-Vaisala period. The upper boundary of the waveguide is formed heated region of the upper atmosphere. In the figure shows mean temperature in the upper atmosphere and wave temperature perturbations in the upper atmosphere produced by a ground-based AGW source. 2 hours after the power switching can be noted a significant change in wave propagation mode is manifested in the character of the waveguide propagation of atmospheric waves in the region below ~ 200 km, and the absence of wave disturbances above 250-300 km.

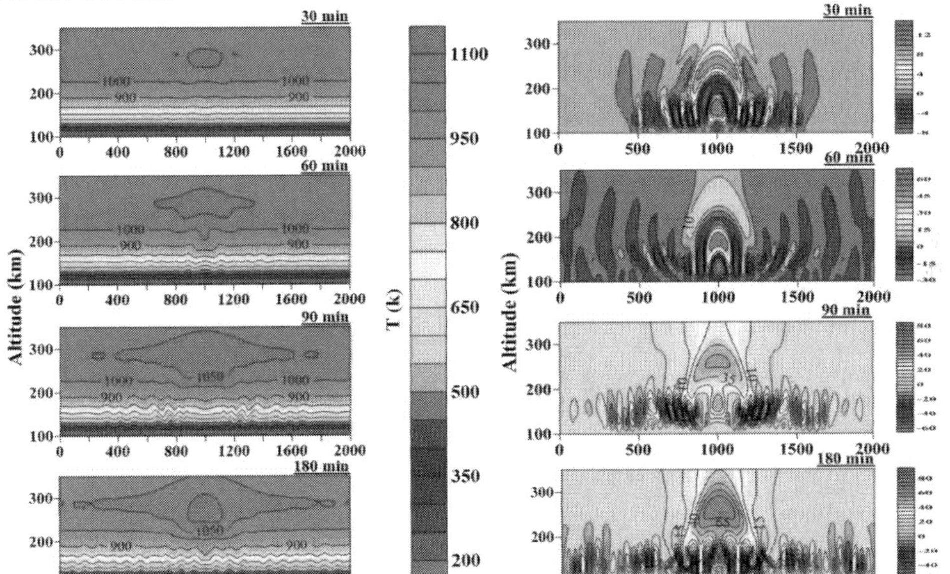

Fig. 1. Mean temperature in the upper atmosphere (left); 30, 60, 90 and 180 min. after start of operation of the ground-based AGW source. Wave temperature perturbations in the upper atmosphere produced by a ground-based AGW source (right) 30, 60, 90 and 180 min. after the start of operation of the ground-based AGW source. The horizontal coordinate of the source is 1000 km;

3 Diurnal variations in atmospheric and ionospheric parameters during a meteorological storm

To examine ionospheric dynamics in conditions of meteorological disturbances, we have selected periods of the strongest meteorological storms in the Kaliningrad region. We analyze ionospheric data, using observations of the F2 layer critical frequency (foF2) and total electron content (TEC) obtained at the station Kaliningrad. Methods of determining foF2 from observations, made by the ionosonde Parus, and TEC by analyzing signals from GNSS satellites are described in [14, 15]. Data on meteorological conditions have been taken from www.rp5.ru.

This paper focuses on rapidly growing meteorological disturbances followed by high winds of force 6–8 on the Beaufort scale. Such storms in the Kaliningrad region generally last less than two days. To identify characteristic features of ionospheric disturbances arising during meteorological disturbances, we consider an observation period that began five days before a storm and ended five days after it. As weather parameters, we analyze diurnal variations in atmospheric pressure P0 at the altitude of the station and maximum gusts of wind at an altitude of 10 m over a 3-hour period between observations — the parameter FF3. To minimize the influence of geomagnetic factors on ionospheric variations, we chose meteorological disturbances under quiet geomagnetic conditions.

Figure 3 illustrates variations in atmospheric and ionospheric parameters during a meteorological disturbance of August 16, 2010. On August 13–23, geomagnetic conditions were quiet; $K_p \leq 3$ (Figure 2). Solar activity changed very little, decreasing from $F10.7=86$ on August 13 to $F10.7=76$ on August 23.

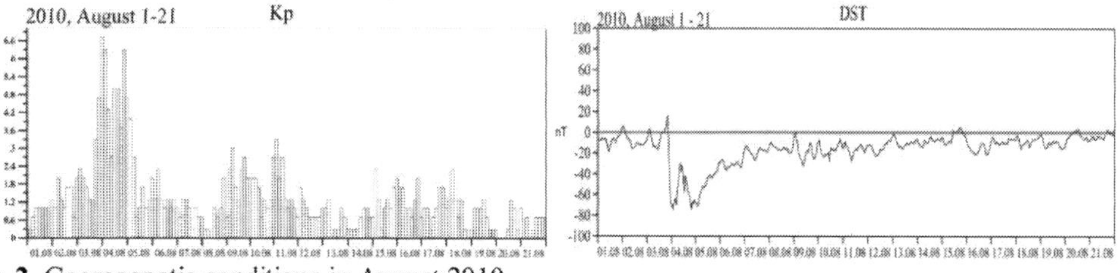

Fig. 2. Geomagnetic conditions in August 2010.

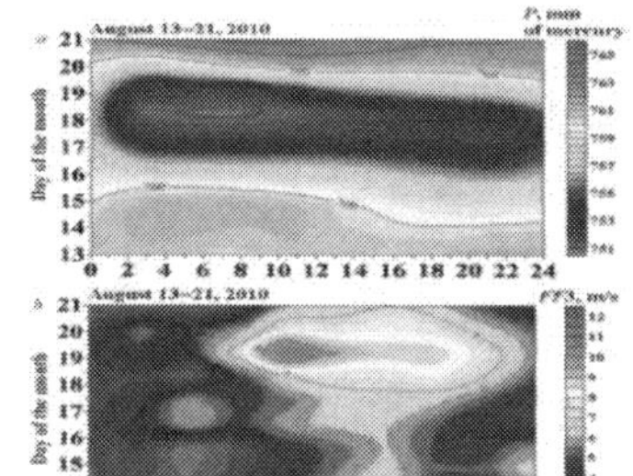

Fig. 3. Diurnal variations in surface air pressure (a); diurnal wind speed variations (b); diurnal TEC variations (c); diurnal foF2 variations (d) on August 13–21, 2010 according to observations from the station Kaliningrad.

It is apparent that atmospheric pressure began to drop on August 14, reached its minimum on August 18, on the day of the storm, and began to rise on August 20 (Figure 3, *a*). The wind speed during the observation period increased largely during daytime on August 18–19, reaching its peak values of ~ 14 m/s on August 19. Figure 3 *c*, *d* shows that TEC and f_0F2 disturbances were pronounced on August 14 and 18 (the white color in Figure 3, *d* indicates observation areas in which we failed to find critical frequency values owing to strong absorption in the E and D regions). Those days, decreases in daytime TEC values run to ~30–40 % and in the critical frequency, to ~ 15–20 % as compared to stormless days.

Similarly, were analyzed another 15 meteorological storms [16]. In the study of some it was observed that during periods of meteorological disturbances occur in the ionosphere diffusivity and F scattering. As can be seen from the figure 4, while weather the storm on August 16 these phenomena have taken place mostly at night time.

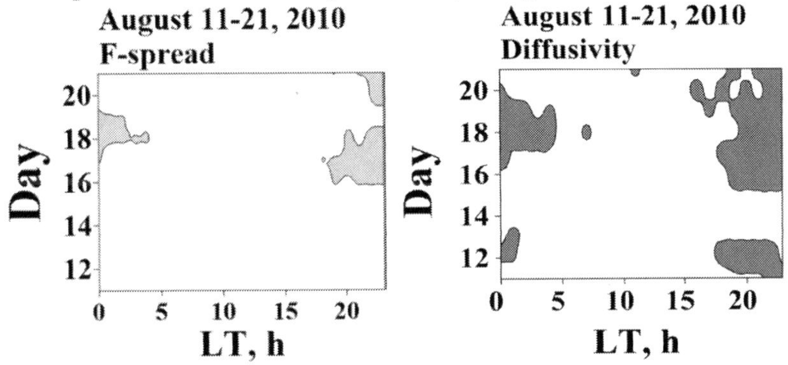

Fig. 4. Phenomena diffusivity and F scattering during periods of meteorological disturbances on August 13–21, 2010 according to observations from the station Kaliningrad.

4 Discussion

We can suppose that rapid changes in meteorological conditions during a storm favour excitation of AGWs over a wide range of periods. Propagation of AGWs to the upper atmosphere and their dissipation there result in the observed ionospheric disturbances. Theoretical studies of AGW propagation from sources in the lower atmosphere suggest that waves of this range can quickly reach the upper atmosphere and due to dissipation can give rise to large-scale disturbances, specifically to local heating regions. Noticeable large-scale disturbances in the upper atmosphere conditioned by dissipation of AGWs propagating from a source in the lower atmosphere are observed 1–2 hr after activation of the source [17]. A local heating region appearing in the upper atmosphere should affect the ionization balance in the ionosphere and lead to a decrease in the ionospheric electron density due to the growing influence of recombination processes.

Figure 5 shows the spectra of variations of the total electron content (TEC, an ionospheric parameter) with periods of 2 to 20 min, changing during observations, over the KLG midlatitude station (54.42° N, 20.27° E). Analysis of TEC results showed that during the period of meteorological storms intensified infrasonic components with periods of 6-10 minutes, lasting for days after the storm.

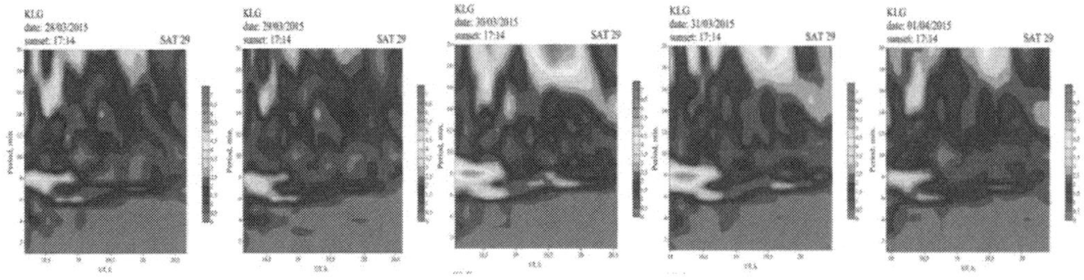

Fig. 5. Analysis of TEC results in signals from some GPS satellites during the period of meteorological storms March 29-30, 2015.

5 Conclusion

1. During the period of meteorological perturbations marked decreasing in the amplitude of diurnal variations of TEC up to ~ 50%; for foF2 - ~ 20%. in relation to meteorologically calm days;
2. During the period of meteorological perturbations observed phenomena of F-scattering and diffuse in F region of the ionosphere;
3. Analysis of spectra of atmospheric variations detects the existence of harmonics with periods of acoustic gravity waves and infrasound waves in the ionosphere during meteorological storm;
4. AGWs, excited by localized sources on the Earth's surface, propagate to the heights of the upper atmosphere and ionosphere. The process of propagation of AGWs accompanied by interrelated effects due to the wave dissipation in the upper atmosphere: heating of the atmosphere by waves, which leads to local changes in atmospheric conditions, and waveguide propagation of wave perturbations with periods of less than a period Vaisala - Brent in the disturbed atmosphere, at altitudes below 200 km;
5. The most probable cause of ionospheric perturbations is the processes of formation of local area heating the thermosphere due to AGWs dissipation coming into the upper atmosphere from the area of meteorological perturbations in the lower atmosphere.

These investigations were performed with financial support of the Russian Foundation for Basic Research Grants No. 15-05-01665 and State task Education and Science Ministry of the Russian Federation, the competitive part of the task No 3.1127.2014/K.

References

1. D. Altadill, E.M. Apostolov, J.G. Solé, Ch. Jacobi, Physics and Chemistry of the Earth, Part C: Solar, Terrestrial & Planetary Science **26**, 6 (2001)
2. P. Sauli, J. Boska, Physics and Chemistry of the Earth, Part C: Solar, Terrestrial & Planetary Science **26**, 6 (2001)
3. J. Lastovicka, P. Sauli, Advances in Space Research. **24**, 11 (1999)
4. D.C. Fritts, M.J. Alexander, Reviews of Geophysics **41**, 1 (2003)
5. M.A. Chernigovskaya, B.G. Shpynev, K.G. Ratovsky, Journal of Atmospheric and Solar-Terrestrial Physics **11**, 22 (2015)
6. P. Sauli, J. Boska, Journal of Atmospheric and Solar-Terrestrial Physics **63**, 9 (2001)
7. S.L. Vadas, H.-L. Liu. Journal of Geophysical Research: Space Physics **114**, A10310 (2009)

8. T. Sindelarova, D. Buresova, J. Chum, F. Hruska, Advances in Space Research **43**, (2009)
9. A.S. Polyakova, N.P. Perevalova, Advances in Space Research (includes Cospar Information Bulletin) **52**, 8 (2013)
10. S.P. Kshevetskii Analytycal and numerical investigation of nonlinear internal gravity waves. Nonlinear processess in geophysics **8**, 37-51 (2001)
11. S. P. Kshevetskii, Computational mathematics and mathematical physics **41**, 12 (2001)
12. S.P. Kshevetskii Computational Mathematics and Mathematical Physics **41**, 2. (2001)
13. I.V. Karpov, S.P. Kshevetsky, O.P. Borchevkina, A.V. Radievsky, A.I. Karpov, Russian Journal of Physical Chemistry B **10**, 1 (2016)
14. A.L. Karpenko, N.I. Manaenkova, Geologishe Rundshau **85**, 1 (1996)
15. L.W. Baran, I.I. Shagimuratov, N.J. Tepenitzina The Use of GPS for Ionospheric Studies. Artificial satellites. Journal of Planetary Geodesy. 1997, vol. 32, no. 1, pp. 49–60;
16. I.V. Karpov, O.P. Borchevkina, R.Z. Dadashev, A.V. Ilminskaya, Solar-Terrestrial Physics **2**, 2 (2016)
17. I. V. Karpov, S. P. Kshevetskii, Geomagnetism and Aeronomy **54**, 4 (2014)
18. http://www.rp5.ru

Earth's electromagnetic field by observations in Buryatia

Gennady Drugin[1,*], *Yury* Bashkuev[2], *Idam* Naguslaeva[2], *Nina* Cherneva[1], and *Boris* Shevtsov[1]

[1] *Institute of Cosmophysical Researches and Radio Wave Propagation FEB RAS, Paratunka, Kamchatsky Kray, Russia*
[2] *Institute of Physical Material Science SBRAS, Ulan-Ude, Russia*

Abstract. The data obtained by the World Wide Lightning Location Network (WWLLN) in the territory of Zabaykal'e are compared with the data received by the equipment for observations of Earth's natural pulse electromagnetic field installed in Buryatia. Diurnal and seasonal dependence of lightning strokes observed in Zabaikal'e territory were obtained. It was shown that the maxima in diurnal variation registered by the instrumentation in Buryatiya coincide with the main world lightning sources.

1 Introduction

Lightning activity is a hazardous phenomenon. Information on lightning may be applied in different areas of human activity. Lightning strokes can cause much damage due to the induced currents in power lines and cable links. Information on lightning is important to provide flight and navigation safety and prevention of forest fires from lightning activity. For example, it was written on [http://www.altesmedia.ru/index.php/component/k2/item/797-grozy-stali-prichinoj-pozharov-v-zabajkalskom-natsionalnom-parke] that in July 2015 ?forest fires in Buryatiya were registered in the territory of Zabaykal'skiy National Park. According to the data of the Republic Forestry Agency, the cause of fires on the territories was dry lightning?

The World Wide Lightning Location Network (WWLLN) [http://wwlln.com] was developed to investigate lightning activity. It includes about 50 sites distributed all over the world. The WWLLN enables us to register lightning locations with good accuracy. However, this network records only powerful lightning strokes. In some regions of the Earth this network is not capable of registering of some lightning due to their remoteness from registration sites and a small number of sites.

In the territory of the Eastern Siberia, instrumental observations of lightning activity are carried out only in some separate regions [1, 2]. Investigations of the radiation associated with the Earth natural pulse electromagnetic field (ENPEMF) are carried out in Zabaykal'e, Buryatiya. However, the registration sites in this territory are not included into the WWLLN network. Thus, at present it is impossible to monitor lightning activity and to determine the location of lightning sources there.

The aim of the paper is to define the main characteristics of radiation from lightning sources based on the data of ENPEMF registration in Buryatiya territory and to show the capabilities of WWLLN for monitoring of lightning activity in Buryatiya.

*e-mail: drug@ikir.ru

2 Instrumentation and methods of investigation

Some results of the investigations and the WWLLN network are described in [3]. Registration of ENPEMF was carried out in Ulan-Ude at the Institute of Physical Material Sciences (IPMS) SB RAS. ENPEMF flux densities have been observed in VLF range by a multichannel geophysical recorder MGR-01 [4] from March 31, 2008 till the present moment [4].

According to the registration data of ENPEMF magnetic components from the north-southern and east-western directions, time dependence of pulse number for 2008-2015 was plotted (Fig. 1). Annual variation of pulse stokes has good repetition. The largest number of pulse strokes was observed in summer and reached 100 000 and more, in winter time the average value of pulses did not exceed several of hundreds.

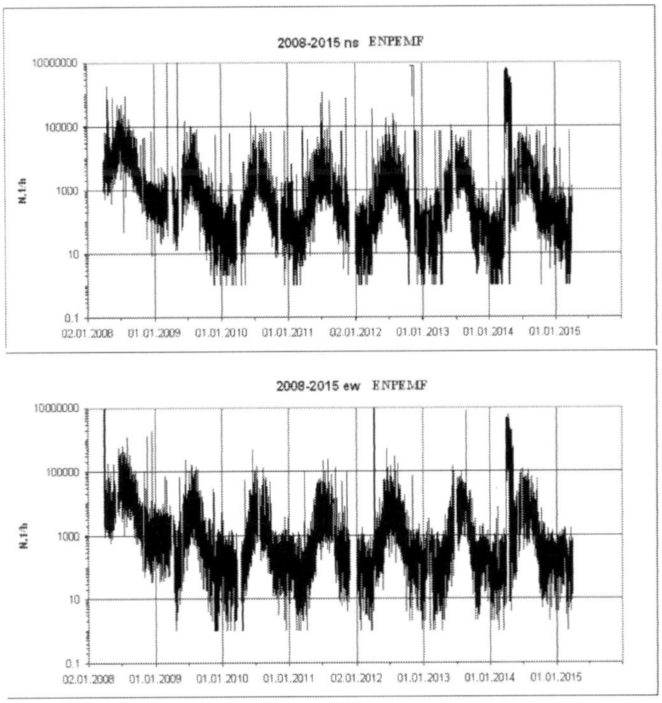

Figure 1. Lightning strokes N, recorded in Buryatiya from the north-southern (n-s) and east-western (e-w) directions for 2008–2015.

3 Annual variation

For detailed consideration of the annual variation, we chose the year of 2013 and plotted the dependence of pulse number recorded by WWLLN and by the instrumentation for ENPEMF from the north-southern and east-western direction (Fig. 2). The number of lighting pulses from WWLLN data was estimated from the region (Fig. 3) limited by the coordinates ($45^0 - 55^0$)N, ($100^0 - 115^0$)E, in the center of which is the ENPEMF registration site, ENPEMF Ulan-Ude (φ=51.83^0, λ=107.62^0).

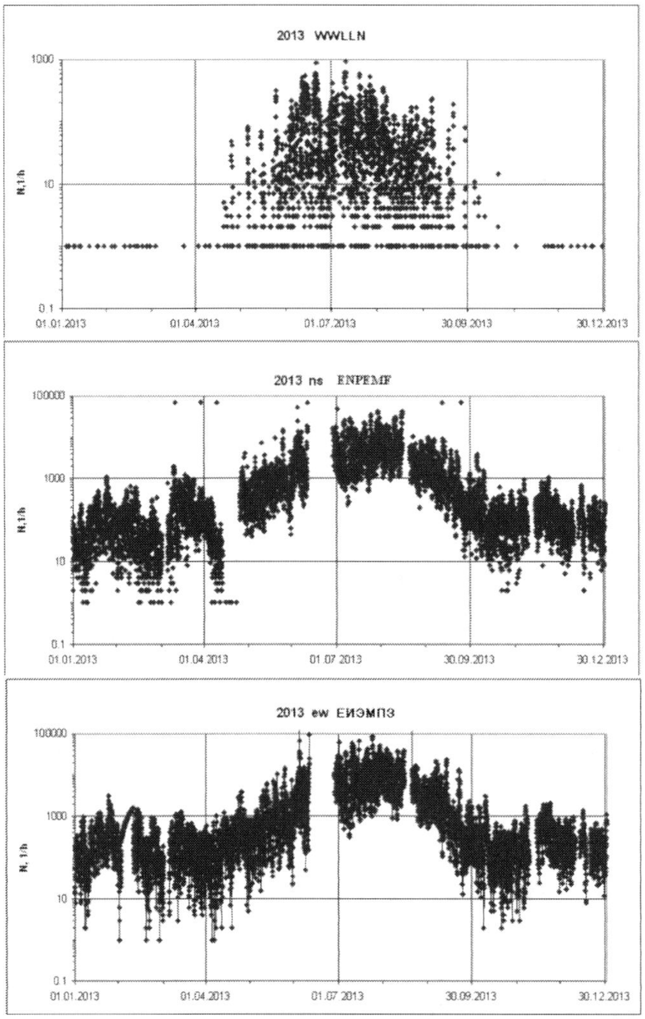

Figure 2. Lightning discharges N, recorded by the WWLLN network (at the top) and ENPEMF instrumentation from the north-southern (n-s) and east-western (e-w) directions for 2013.

It is clear from Fig. 3 that according to WWLLN, the number of lightning was ~ (1 – 1000) 1/hour, and the pulse number recorded by ENPEMF instrumentation was ENPEMF ~ (1 – 100000) 1/hour. That is about two orders as much compared to the WWLLN network.

The difference in the number of recorded radiations between WWLLN and ENPEMF data may be explained by the fact that in winter time, local lightning occurs rarely and WWLLN does not detect them. However, the ENPEMF instrumentation records local and remote lightning as well as

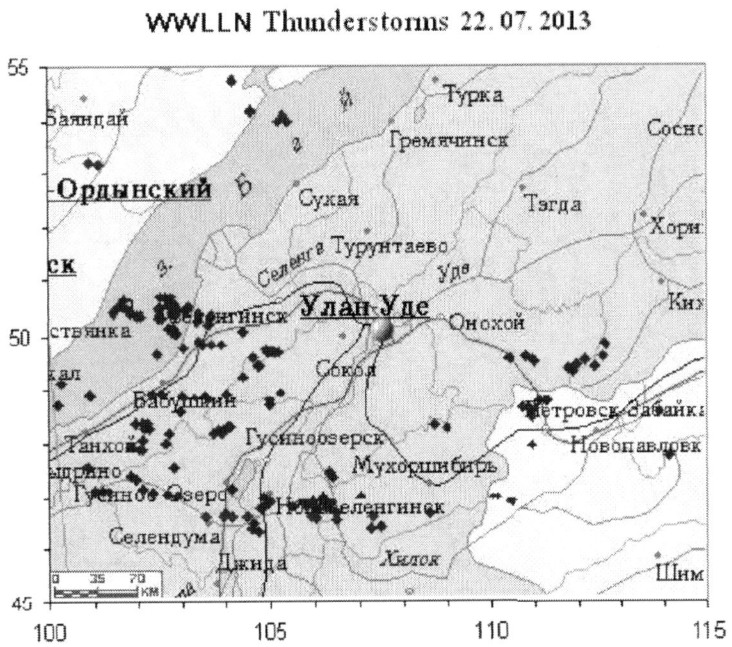

Figure 3. Zabaikal'e territory chosen to estimate the number of lightning. Squares are lightning strokes occurred on June 22, 2013.

the radiations from the world lightning sources located in the near equatorial regions of the Earth [5]. Lightning from the world sources had been registered before in the north-eastern regions of Russia which are even more distant from the equator [6].

4 Diurnal variation

The diurnal variations of the averaged monthly pulses recorded by the WWLLN network and by the ENPEMF site are shown in Fig. 4. The bottom part is the diurnal variation of the world lightning activity [5].

It is clear (Fig. 4,1) that in January the WWLLN recorded just a few number of lightning. In winter time, local lightning is not almost observed and there is no diurnal variation. However, quite a large number of them was recorded by the ENPEMF instrumentation with a clear diurnal variation. Its maximum value was observed at 20 h UTC (Fig. 4,2-3) that almost coincides with the maximum of lightning activity of the American lightning source (Fig. 4,14-15). Moreover, the larger number of pulses arrived from the east-western direction than from the north-southern direction. That also confirms the reception of pulses from the American lightning source which is located in the eastern

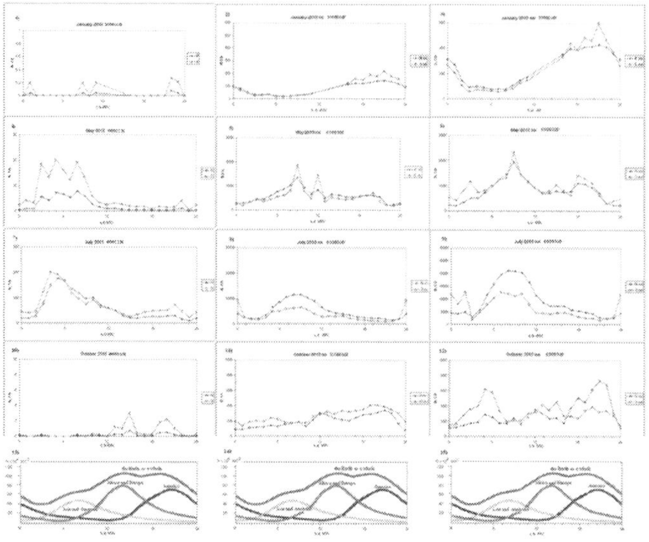

Figure 4. Diurnal variation of averaged monthly pulses received by the WWLLN network and ENPEMF station in January, May, July, November 2013 (N ? average values, S ? root mean square deviations). At the bottom are the diurnal variations of the world lightning activity.

direction from Buryatiya. Another maximum, less in values, is at 0 h UTC. It also coincides with the less maximum of lightning activity in the American lightning source.

In May the number of lightning recorded by the WWLLN network significantly increased, the diurnal variation is clearly observed with the maximum at 6 h UTC (Fig. 4,4). That corresponded to 16 h LT, local afternoon time in Buryatiya. From the ENPEMF data, the maximum value was at 9 h UTC (Fig. 4,5-6), local evening time, 19 h LT. This discordance of occurrences of the maxima may be explained by the fact that the WWLLN network records the radiation from lightning in a definite region (45^0 – 55^0)N, (100^0 – 115^0)E, whereas the ENPEMF instrumentation also receives radiation out of this region, mainly from the south-eastern direction. In May, there is also the second maximum in ENPEMF at 19 h UTC, which is less in value and which almost coincides with the maximum observed in the American lightning source in January.

The annual maximum of lightning activity is observed in summer time, in July. The WWLLN network recorded the maximum number of lightning at 5 h UTC (Fig. 4,7). The maximum of ENPEMF radiation is at 8 h UTC (Fig. 4,8-9). On this basis we can make a conclusion that the ENPEMF instrumentation recorded radiations outside the chosen region, supposedly from the south directions. We should also note that 8 h UTC almost coincides with the time of maximum activity of the Asian and Australian lightning source (Fig. 4,14-15) which is the closest one to the observation site.

In October the WWLLN network recorded a small number of lightning in the defined region (Fig. 3) and the diurnal variation was not clearly observed (Fig. 4,10) whereas the diurnal variation obtained by the ENPEMF instrumentation was clear (Fig. 4,11-12). According to the ENPEMF data, three maxima were observed. Each of them almost coincided in time with a corresponding world lightning source (Fig. 4,14-15). We may suppose that the ENPEMF instrumentation mainly recorded remote lightning in October.

5 Conclusions

1. Measurements of the number of lightning strokes recorded by ENPEMF instrumentation in 2008–2015 in Buryatiya territory have been carried out. Diurnal and seasonal dependencies of the number of lightning strokes in Zabaikal'e territory for 2013 have been received applying the WWLLN network as well as the dependences registered by the ENPEMF instrumentation.

2. The comparative data analysis has been carried out applying the WWLLN network and the ENPEMF instrumentation recording radiations in Zabaikal'e territory. The analysis has showed that the observed number of lightning during a year changes within the range of (1 ? 1000)1/h, and the ENPEMF instrumentation receives pulses from lightning strokes by two orders more. The difference in number is explained by the fact that the ENPEMF instrumentation received radiations both from local and from remote sources including the world lightning centers.

3. The radiation diurnal variations from WWLLN and ENPEMF data have been compared with the diurnal variations of the African (Africa and Europe), Asian (Asia and Australia) and American lightning sources. It has been obtained that there are maxima coinciding with the world lightning centers in the diurnal variations recorded by ENPEMF instrumentation in Zabaikal'e territory.

4. Installation of WWLLN instrumentation in Buryatiya territory will allow to record lighting sources in Zabaikal'e territory with more accuracy and to respond promptly on negative processes occurring in the result of lightning activity effect.

References

[1] V.I. Kozlov, V.A. Mullayarov *Thunderstorm activity in Yakutia* (Pub.YB SB RAS, Yakutsk, 2004), 104

[2] G.I. Druzhin, N.V. Cherneva, A.N. Mel'nikov, Geomagnetizm and Aeronomy, **49:8**, 1305–1307 (2009)

[3] R.H. Holzworth et al., Geophysical Research Letters, **31:3**, 5 (2004)

[4] E.L. Afraimovichetal et al., *Seismic-ionosphericand seismic-electromagnetic processes in the Baikal rift zone* (House of the SB RAS, Novosibirsk, 2012), 304

[5] *Handbook of geophysics* (Nauka, Moscow, 1965), 571

[6] G.I. Druzhin, V.I. Shapaev, Geomagnetism and Aeronomy, **28:1**, 81–86 (1988)

Relation of daily periods of vlf radiation with X-ray sources

Gennady Drugin[1,*] and *Ivan* Stasiy[1]

[1] *Institute of Cosmophysical Researches and Radio Wave Propagation FEB RAS, Paratunka, Kamchatsky Kray, Russia*

Abstract. Spectral analysis of electromagnetic noise radiation in the VLF range at three fixed frequencies for the period from 1997 to 2015 have been carried out, Periodograms with diurnal components associated with the periods of the Earth rotation relatively the Sun and the stars have been obtained. It has been shown that the Sun x-radiation and the galaxy x-radiation affect daily variations of VLF radiation.

1 Introduction

The Sun x-radiation is the major factor affecting the ionosphere D region http://ru.wikipedia.org/, [1, 2]. Additional weak sources of ionization such as meteorites burning up in the atmosphere at the heights of 60-100 km, cosmic rays, energetic particles of the magnetosphere brought into this layer during magnetic storms also have some impact on the D region ionization process which mainly affect the conditions of radio wave propagation in a very low frequency range.

Sun daily rotation (~ 27 days) as well as the week cycle of human activity may also manifest themselves in the spectra of electromagnetic VLF radiation received on the Earth [3]. However, there are only single cases when daily periods of VLF radiation are under consideration [4, 5].

The aim of the paper is the investigation of noise characteristics of electromagnetic VLF signals in the period ranges close to the Earth daily rotation on large time intervals and determination of their potential relation with x-radiation sources.

In order to do that, we have carried out spectral analysis of VLF radiation noise envelope at three fixed frequencies based on the data of long-term registration in the period range close to the Earth daily rotation.

2 Instrumentation and methods of investigation

Electromagnetic noise radiation was registered by a multichannel VLF-detector located in Kamchatka at «Karymshina» field site (φ = 52°49′ N, λ = 158°07′ E) The VLF radiation envelope is recorded at three fixed frequencies in narrow frequency bands (5% - 10% from the band center). The VLF-detector consists of a loop antenna, pre-amplifier, communication cable, signal filtration unit, output device. The loop antenna has a size of 7,5x15 m and its plane is oriented in the east-western direction.

Spectral analysis was carried out for the radiation envelope at the frequencies of 0.7, 1.2, 5.3 kHz near the daily periods. Fig. 1,a shows a periodogram for the registration period from 1997 to 2006

* e-mail: drug@ikir.ru

(10 years). It is clear from the figure that there is the main maximum at the period of 1440 min (1 day) at all the frequencies. This period is associated with the Earth daily rotation relatively the Sun. Moreover, another maximum less in value is observed at the period of 1436 min. This period coincides with the period of Earth rotation relatively remote stars. Periodograms for the registration period indicated above were illustrated in the paper [4, 5].

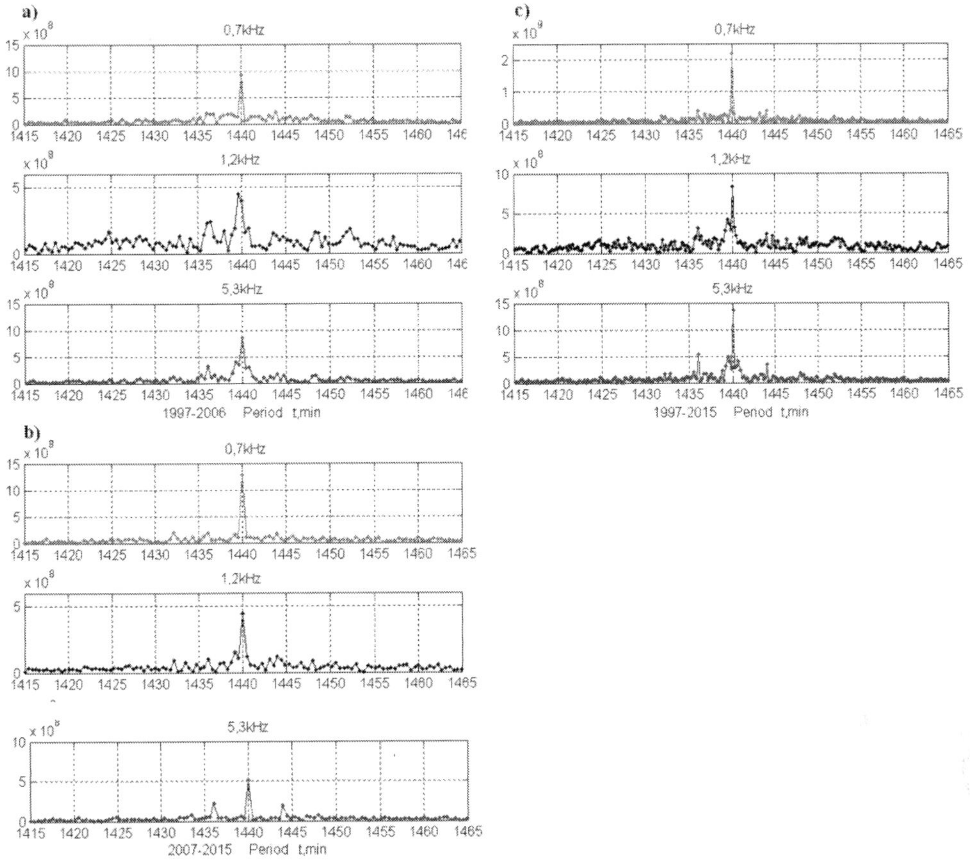

Figure 1. eriodograms of VLF noise envelope at three frequencies near daily periods based on observations for a) 1997 – 2006 b) 2007 – 2015 c) 1997 – 2015.

Further analysis of the radiation envelope near the daily periods at the same frequencies was carried out for the period from 2007 to 2015 (9 years). It is clear from Fig. 1,b that there are main and additional maxima at the periods of 1440 and 1436 min during the observation period. There is another maximum at the period of 1444 min which is more distinct at the frequency of 5.3 kHz. This maximum does not coincide with daily periods of the Earth rotation relatively the Sun and the stars. It is most likely to be determined by the annual rotation of the Earth around the Sun since the daily variation of the VLF radiation envelope at the frequency f_0 is modulated by the annual variation with the frequency Δf. Thus, due to the annual variation, two components similar in amplitude at the frequencies of $f_1 = f_0 + \Delta f$ and $f_2 = f_0 - \Delta f$, (Δf is the modulation frequency)should appear in a signal spectrum besides the fundamental frequency of f_0. In our case, the fundamental period is $T_0 = 1/f_0 = 1440$ min, $T_1 = 1/f_1 = 1436$ min, $T_2 = 1/f_2 = 1444$ min. Only the period T_1 coincides with the «star» one out of these «annual» periods.

Fig. 1,c shows a periodogram for the registration period from 1997 to 2015 (19 years). The merged data sets (1997 – 2006 and 2007 – 2015) for spectral analysis allowed us to improve the oscillatory period resolution. On this periodogram, maxima at the periods of 1440 and 1436 min are more distinguished. There are maxima at the periods of 1444 min but they are less in amplitude.

Now we consider the dependence of the detected maxima on observation periods, noise and registration frequency. We specify the natural noise radiation as a signal, and the instrumentation self-noises and different other kinds of disturbances as noise. The average of noise is calculated for the spectral component range from 1415 to 1465 min excluding the period from 1435 to 1445 within which the signals associated with daily and annual variation of natural electromagnetic radiation. Then signal amplitude values at the periods of T_1, T_0 and T_2 are divided by the noise average and the results are written into the table.

Tables 1 – 3 show spectral component K values divided by the noise averages for different time intervals of VLF radiation for the envelopes at three frequencies.

Table 1

1997 – 2006			
T (min)	1436	1440	1444
K (0.7 kHz)	4.5	20.8	4.6
K (1.2 kHz)	3.2	5.7	1.8
K (5.3 kHz)	7.9	21.4	4.2

Table 2

2007 – 2015			
T (min)	1436	1440	1444
K (0.7 kHz)	3.9	26.9	3.6
K (1.2 kHz)	3.1	13.6	3.6
K (5.3 kHz)	9.2	21.0	4.9

Table 3

1997 – 2015			
T (min)	1436	1440	1444
K (0.7 kHz)	5.8	32.7	5.6
K (1.2 kHz)	4.0	10.7	3.1
K (5.3 kHz)	11.4	28.7	7.3

It is clear from Tables 1 – 3 that K main daily maximum at the period of $T_0 = 1440$ min significantly exceeds the noise level as well as K maxima at the periods of $T_1 = 1436$ min and $T_2 = 1444$ min at all the registered frequencies. For the most part, K maxima at the periods of 1436 min and 1444 min also significantly exceed the noise level.

For the observation period of 1997-2006 (Table 1), K values at the frequency of 0.7 kHz at the periods of 1436 min and 1444 min are almost the same and exceed the noise level by more than 4 times. It proves that at this frequency the K maxima at the periods of 1436 min and 1444 min are generally determined by radiation annual variation, and a small difference in the values is due to the noise effect.

At the frequency of 1.2 kHz, the difference between K at the periods of 1436 min and 1444 min (Table 1) is more than that considered above that indicates the presence of an additional "star" type source of radiation besides the annual one. We should note that at this frequency the K values are less than at other frequencies that means that the relative impact of noise is higher.

At the frequency of 5.3 kHz, the K values are the highest for all the periods T shown in Table 1. Both "star" and "annual" periods are clearly distinguished at this frequency.

Almost everything that has been described for the observation period from 1997 to 2006 is true for the period from 2007 to 2015 excepting that at the frequency of 1.2 kHz the K values at the periods of 1436 min and 1444 min are almost similar. "Annual" period appears at this frequency and a small exceedance of K at the period of 1444 min over the period of 1436 min is explained by noise effect.

K values for the whole period of observations from 1997 to 2015 (Table 3) show that with observation time increment, the signal/noise relation increases, and "star" period is distinguished more clearly. "Annual" periods are also distinctly observed, for example, at the frequency of 0.7 kHz where the difference between K values at the periods of 1436 min and 1444 min is small. Just like in the previous cases, "star" period of 1436 min at the frequency of 5.3 kHz is clearly observed.

3 Sources of radiation

It is known that the main sources of natural electromagnetic radiation in VLF range are lightning strokes. When such radiation is received, the level of a signal depends on the intensity of a lightning source and the conditions of signal propagation along the path. In summer time, radiation from nearby (regional) lightning sources is mainly registered in the North-East of Russia. During this period, the level of the received noise is maximal. In winter time, there is no local lightning, and the VLF receiver registers remote lightning located in the near equatorial regions for the most part. Radiation from the world lightning centers is recorded [6]. During this time, the level of the signal depends significantly on the conditions of signal propagation along the path. At nighttime radiation generated by the sources at the distances of 10 000 km from the registration site is registered.

When radiation propagates from remote sources, the main factor affecting registration of VLF signals is the Sun x-radiation. In the paper [1] the authors found a close relation of VLF noise intensity variations with x-radiation flux changes during solar flares. Whereas, the paper [2] considers the method of estimation of x-radiation fluxes from solar flares by the level of absorption of regular noise VLF radiation. The intensity of solar flare x-radiation may reach high values. For example, the maximum intensity of a flare which occurred on November 4, 2004 was $28 \cdot 10^{-4}$ W/m^2 [http://ru.wikipedia.org/wiki/]. The capability of x-ray flares to cause sudden ionospheric disturbances depends both on the flux value and on the degree of stiffening during a flare estimated as a relation of flux levels in spectral intervals $0 - 3\ A^0$ and $0 - 8\ A^0$ [7]. Besides the solar radiation, there is also galactic radiation in the x-radiation range which may also influence the level of the registered VLF noise affecting the D and E regions of the ionosphere. S.L. Mandelshtam wrote [http://w-rabbit.narod.ru/raznoe/rentgen.htm], hat by the Earth's atmosphere boundary, the Sun x-radiation flux in the region of 1 – 10 angstrom is 10^{-4} erg/(sm^{2}s) (total flux of electromagnetic radiation from the Sun by the Earth's atmosphere boundary is $1.5 \cdot 10^6$ erg/(sm^{2}s), and the most powerful source of the Galaxy is in Scorpio constellation Sco X-1, the flux from which is $5 \cdot 10^{-7}$ erg/(sm^{2}s) (from the galaxy, $10^{-4} - 10^{-3}$ of optical luminosity fall on the x-ray range). There are about a hundred of discrete x-ray sources. The x-radiation is also observed from the Galactic ridge x-ray emission. It is extensive radiation of low surface brightness arranged as a band with the width of about 1 – 2 degrees along the Galaxy plane [http://ru.wikipedia.org/wiki/]. Moreover, based on reasonable, statistically valid estimates, we can expect about a hundred of metagalactic x-ray sources [http://w-rabbit.narod.ru/raznoe/rentgen.htm],but the flux from these sources coming to the Earth's atmosphere is small due to the great distance. It follows from the above that the x-radiation flux from the Galaxy differs in value from that from the Sun by the Earth's atmosphere boundary. It also may affect the level of ionization of the ionosphere D and E regions that causes the change of the registered VLF radiation amplitude.

Besides the x-radiation, ionospheric and magnetospheric storms as well as industrial noises, circuit noises of 50 Hz and its harmonics may affect registration of noise VLF signals. Ionospheric-magnetospheric storms make the major impact during the registration of VLF signals at high latitudes, in the auroral belt. Industrial noises are mainly observed near towns and manufacturing areas. In our case, the observation site is located far from the auroral belt and from noises of industrial origin. However, there is some small impact of the mentioned sources on the noise component of VLF signals.

X-radiation influencing the process of ionosphere ionization affects the frequency characteristics of the registered VLF signals in different ways. It was noticed that the radiation noise level usually increased in the frequency range less than 1 kHz during solar x-ray flares, whereas it significantly decreased at the frequencies of 4-5 kHz. In our case, the registered radiation at 0.7 kHz and 1.2 kHz is in the transition frequency region which is weakly affected by the x-radiation. Thus, the impact of x-ray sources at these frequencies is minimal. Almost similar K values at the periods of 1436 min and 1444 min at these frequencies (Tables 1 – 3) indicate that "annual" impact is greater than the "star" one.

The frequency of 5.3 kHz falls within the range with maximum influence of x-radiation, thus, the effect should be more significant at this frequency that is observed in K values (Tables 1 – 3). Considerable exceedance of K at the period of 1436 in comparison to the period of 1444 shows that the "star" source prevails over the "annual" one.

4 Conclusions

We have carried out a spectral analysis of noise electromagnetic radiation envelope at the frequencies of 0.7, 1.2, 5.3 kHz near the daily periods during the observations from 1997 to 2006 (10 years), from 2006 to 2015 (9 years) and for the whole period of observations from 1997 to 2015 (19 years).

Radiation maxima are observed in spectral components at the periods of 1436, 1440 and 1444 min. All the recorded maxima for 1997-2015 exceed the average noise by more than 3 times. The maximum exceedance of the radiation over noise is 28.7, and it is observed at the period of 1440 min for the received frequency of 5.3 kHz.

The spectral component at the period of 1440 min is determined by the Earth daily rotation relatively the Sun, whereas at the period of 1436 it is determined by the Earth daily rotation relatively the stars. There are also annual variations at the periods of 1436 and 1444 min in the spectral components of noise electromagnetic radiation.

It has been shown that the Galaxy x-radiation may affect the ionization level of ionosphere D region and its influence causes a maximum in noise VLF radiation spectrum at the period of 1436 min associated with the Earth rotation relatively the stars. The obtained result may promote better understanding of the processes affecting the D region of the ionosphere.

References

[1] N. N. Murzaeva, *Relation of VLF radiation of the upper atmosphere with other geophysical phenomena* (YB SB AS USSR, Yakutsk, 1977), 124

[2] V. A. Mullayarov, R. R. Karimov, V. I. Kozlov, N. N. Murzaeva, Meteorology and Hydrology, **8**, 48–56 (1998)

[3] V. I. Kozlov, V. A. Mullayarov, *Lightning activity in Yakutiya* (YB SB RAS, Yakutsk, 2004), 104

[4] G. I. Druzhin, A. N. Mel'nikov, N. V. Cherneva, Doklady Earth Sciences, **457:1**, 92–94 (2014)

[5] G. I. Druzhin, International research journal, **8:39**, 98–101 (2015)

[6] G. I. Druzhin, V. I. Shapaev, Geomagnetism and Aeronomy, **28:1**, 81–86 (1988)
[7] A. Mitra, *Effects of solar flares on Earth's ionosphere* (Mir, Moscow, 1977), 370

Overhauser vector magnetometer POS-4: Results of continuous measurements during 2015–2016 at geophysical observatory "Paratunka" of IKIR FEB RAS, Kamchatka, Russia

Sergey Khomutov[1,⋆], *Vladimir* Sapunov[2], *Aleksey* Denisov[2], *Dmitry* Savelyev[2], and *Ilkham* Babakhanov[1]

[1] *Institute of Cosmophysical Researches and Radio Wave Propagation FEB RAS, Paratunka, Kamchatsky Kray, Russia*
[2] *Ural Federal University, Quantum magnetometry laboratory, Ekaterinburg, Russia*

Abstract. The results of measurements by Overhauser vector magnetometer POS-4 carried out for one year and a half are presented. The accuracy, stability and reliability are evaluated.

1 Introduction

Continuous observations of the Earth's magnetic field are carried out for characteristic times from seconds to tens of years at magnetic observatories (MO). To provide proper results in such a wide frequency range, two types of devices are used at MOs of INTERMAGNET network, they are: (1) variometers ensuring high frequency and sensitivity but measuring only variations and/or and having low long-term stability, and (2) magnetometers measuring the magnetic field in the absolute sense (full vector) but slow and requiring manual operation in general. Joint application of magnetometers of these types allow us to obtain acceptable results corresponding to INTERMAGNET standards [1]. Modern materials, new digital technologies and processing systems raise a question on the development of an automatic device providing the functions of both variation and absolute magnetometers. A high-performance absolute magnetometer AutoDIF was developed at the Meteorological Institute (Belgium, [2]) for almost continuous measurements of declination and inclination. Scalar magnetometers, Overhauser POS-1 or GSM-19W, are absolute devices to measure the total intensity of field. However, the problem of absolute measurements of field vector has not been solved yet. One of the attempts is the development of vector magnetometer POS-4 (Quantum magnetometry laboratory, Ural Federal University, Ekaterinburg, http://magnetometer.ur.ru).

2 Instrumentation, conditions of measurements

Vector magnetometer POS-4 is a new development; partially, technical and methodical questions of POS-4 were solved in a two-component magnetometer POS-3 [3]. POS-4 enables us to measure the

⋆e-mail: khomutov@ikir.ru

 DOI: 10.1051/e3sconf/20161100007

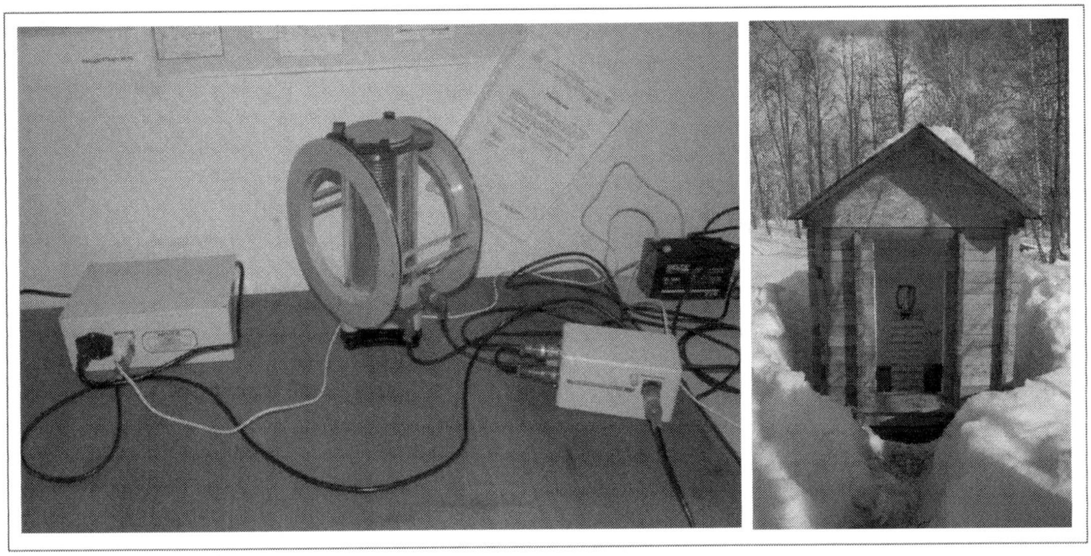

Figure 1. Vector magnetometer POS-4. Left panel: a UPS is to the left, sensor (solenoid and ring system on the carrier) is in the center, electronic unit is to the right. Right panel: POS-4 hut, magnetometer sensor is mounted on a pillar made of building glass stones, the removable hut door is on brass bolts (March 2016)

magnetic induction F and two orthogonal components, the vertical Z component along the Garret solenoid axis and the horizontal projection of vector F on the axis of Helmholtz double-ring system. Moreover, the measured F and Z allow us to estimate horizontal intensity H. The magnetometer includes a measurement unit POS-1, high-stable current source and magnetic systems generating homogeneous switchable magnetic field:

- based on Garret solenoid with titanic frame, the vertical axis of which is adjusted according to liquid levels with scale factor up to 10" to measure vertical component Z;

- based on titanic ring system perpendicular to Garret solenoid axis to measure F vector horizontal projection on ring axis in a random azimuth.

A nonmagnetic base (carrier) with a horizontal scale is used to mount the magnetometer on a pillar. The communication with a PC is provided by RS232 protocol. Standard software POS Manager is used for operation and recording of results. Sensitivity of total intensity is about 0.03-0.05 nT, that of ZH-components is about 0.15-0.30 nT, the module absolute error is not more than 1 nT, that of the components is 3-10 nT, measurement cycle is from 1 to 5 s. To some extent, POS-4 is an analogue of magnetometer dIdD GSM-19FD of GEM Systems (Canada). However, it allows us to estimate in absolute sense not only total intensity F, but the vertical Z component as well. We can also estimate one of the horizontal components (X/Y) when the ring system is oriented accurately. The device picture is shown in Fig. 1 (left panel).

IKIR FEB RAS purchased the magnetometer POS-4 in December 2014 within the framework of RFBR Grant [4]. One of the tasks was to evaluate the possibility of change to the instrumentation by Russian producers. POS-4 is rather an experimental device that requires different works (adjusting, control, testing) coordinated with the developer. Moreover, additional magnetic fields generated by the solenoid and the ring system affect the magnetometers located nearby. That is why a special nonmagnetic hut was built for POS-4. It is made of wooden balks, has the size of about 2×2 m, and insulated (10 cm expanded polystyrene) from the inside. The pillar is made of building glass stones on a nonmagnetic base embedded into the ground by 1 m and has the section of 40×40 cm^2. A glass plate is glued on the upper part. To reduce the possible noises, there is no 220 V power supply in the hut and active temperature stabilization is not provided. However, the temperature inside is being constantly measured by digital temperature sensors at three height levels. The general view of the hut and arrangement of POS-4 sensor are shown in Fig. 1 (right panel). The recorded laptop and the UPS are located in a technical hut at the distance of about 20 m. A standard communication and power cables are laid on the ground surface. The system is grounded on a ground stake near the technical hut. The magnetometer is not GPS-synchronized, UTC reference is realized via the laptop system time during measurement startup. The inner precision generator of the system provided timer stability for a month and more.

In April-May 2015, the magnetometer was tested in different operation modes with different orientation of the ring system (with the axis to the magnetic North, crosswise, according to geographic directions). In June the sensor was finally oriented with the axis of the ring system to the geographic "East-West". The POS Manager software records the results of measurements of five modules (with different polarity of magnetic fields generated by the solenoid and the ring system and directly measured), the date, time, measurement quality parameter QMC, as well as the estimated values of components Z and X,Y (X is the horizontal projection of vector F on the ring system axis). Further, we used only five initial values of the modules. Field components were calculated during the processing without application of any cleaning, smoothing etc. The field elements obtained by POS-4 directly and independently are F, Z (projection on the solenoid axis), Y (projection of the ring system axis). We assume all the components to be measured at the same moment (total cycle of 5 s). Other components were calculated by the obtained F,Z,Y.

3 Measurement results

Regular measurements of the magnetic field by POS-4 began in June 2015. Just like for other observatory magnetometers, certain requirements have been carried out for POS-4: logbook of the operations with the device, processing logs, device photography. Processing of the results is similar to the usuall processing for other magnetometers and includes the following: estimation of baseline values (comparison with the results of absolute observations by standard DI-magnetometer LEMI-203 and a scalar magnetometer POS-1), calculation of minute values and output into standard format files, daily comparison with data of the main variometer at the observatory.

Fig. 2 (left panel) shows daily baseline values of POS-4 (differences between the POS-4 data and the results of absolute measurements after smoothing by approximating spine). Jumps in POS-4 record during adjusting of the sensor were removed (but shown for Y_0 as an example). The average of F_0 is fully explained by spatial gradient of F between POS-4 and absolute hut. Significant constant Y_0 on the undisturbed level is result of inaccurate orientation of the ring system to the geographic East.

It is clear from Fig. 2 that when F gradient stability is high, baselines Z_0, Y_0 show significant oscillations from December 2015, the change rate is 2-3 nT/day. It is likely to be caused by instability of the pillar for POS-4 due to soil freezing. Movement of the pillar is confirmed by the readings of levels at the of POS-4 sensor, the changes reached several divisions, that is up to 1'. Fig. 2 also shows average daily outdoor temperature, recorded by a weather station. We see that at the end of November, the temperature did not exceed 0^oC that could cause soil freezing around the pillar taking into account a low level of snow cover during the winter 2015-2016. The situation is even more complicated because the hut is not heated and air temperature changes significantly differ from the outside temperature, and the snow thickness around the hut is uneven. Significant influence during the freezing may be caused by insufficient ruggedness of the base when the weight of the pillar upper part

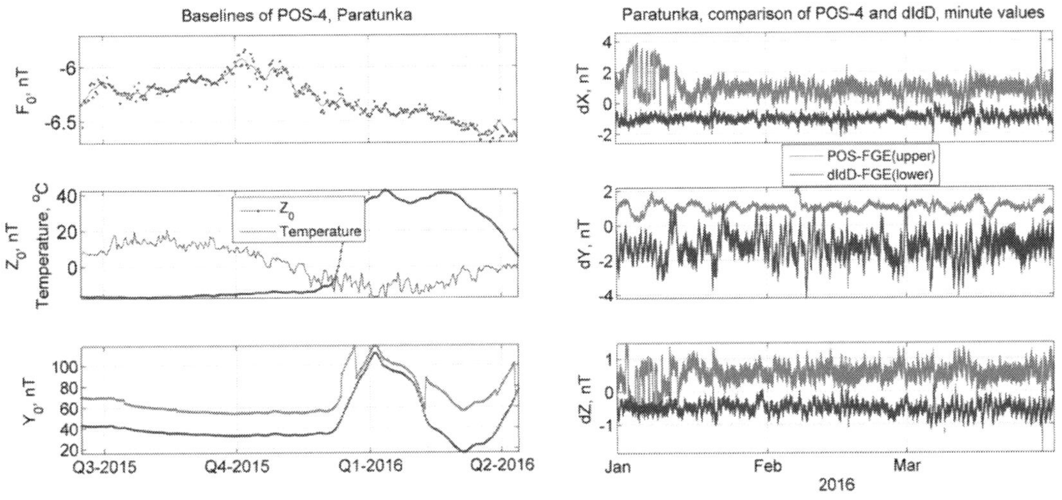

Figure 2. Left panel: adopted (smoothing spline approximation) baseline values of POS-4. Dots show daily mean of F_0, the values of Y_0 are also shown before removal of the jumps during sensor leveling. The average daily outdoor temperature is presented at Z_0 panel. Right panel: comparison of minute data of POS-4 (top curves) and dIdD (low curves). Geomagnetic variations are subtracted (using of FGE magnetometer data). Curves are vertically shifted for better presentation.

made of glass stones is heavy. However, movement of the sensor itself (solenoid or the ring system) may also take place in the result of imperfect supporting base of the carrier or instability of additional field generated by the solenoid or the ring system. We may also note the clear drift acceleration Y_0) after a strong earthquake on January 30, 2016 which occurred at the epicentral distance of about 100 km from the observatory.

The variometer the POS-4 has an analogue, magnetometer dIdD GSM-19FD. Such magnetometers run in continuous mode at IKIR observatories. Fig. 2 (right panel) show minute value sets of both magnetometers for January-March 2016. Due to the strong trend in baseline values of POS-4, their interpolation was made for each minute. Geomagnetic variations were subtracted using the data of fluxgate magnetometer FGE, the main one at the observatory. We may note the following: (1) random components in dIdD and POS-4 data are comparable, it is somewhat higher for X- and Z-components of POS-4 and notably less for Y that is associated with the orientation of the primary sensor of POS-1 in the solenoid; (2) in Y(dIdD) significant variation is observed, it is correlated with temperature (the reason was not defined); (3) jumps in X,Z(POS) in January are caused by air temperature in the hut, however, the mechanism of the effect is not clear; small jumps are caused by adjustment of solenoid verticality.

4 Conclusions

Regular continuous measurements of the magnetic field by the vector magnetometer POS-4 at "Paratunka" observatory in 2015-2016 showed:

1) strong drifts of 2-3 nT/day and more are observed in the data of POS-4. They are likely to be associated with the instability of the pillar in the result of soil freezing and defrosting. Absence of an

automatically compensating system, similar to gimbal suspension of dIdD or FGE, requires a digital system of control for solenoid movement and temperature variations;

2) there is increased noise in Z-measurements of POS-4 (and its derivatives H,X) associated with the orientation of the primary sensor POS-1 in the solenoid that probably requires some change in the construction;

3) more reliable and effective construction for sensor mounting on the pillar is needed, it will provide better stability and accuracy of adjustment along the vertical axis;

4) being more compact and more simple in installation, POS-4 is comparable with the analog dIdD GSM-19FD as a variation device;

5) within more than a year of operation, POS-4 did not have significant problems both in registering hardware and software. However, lack of GPS-synchronization was inconvenient when adjusting the device.

This work was supported by the Russian Science Foundation Grant No. 14-11-00194.

References

[1] St-L. Benoit (Eds.), *INTERMAGNET technical reference manual. Version 4.6* (INTERMAGNET, Edinburgh, 2012), 92

[2] A. Gonsette, A. Poncelet, J.-L. Marin, S. Bracke, J. Rasson, *XVI IAGA Workshop on geomagnetic observatory instruments, data acquisition and processing* (Hyderabad, India, 2014), 29

[3] P. Borodin et al., *Polyarnaya geofizika Yamala: nablyudeniya, bazy dannykh i informatsionnye sistemy v praktike osvoeniya mestorozhdeniy nefti i gaza* (Salekhard, 2014) (in Russian)

[4] S. Yu. Khomutov et al., *Problemy kompleksnogo geofizicheskogo monitoringa Dal'nego Vostoka Rossii* (GS RAN, Petropavlovsk-Kamchatskiy, 2015), 269–273 (in Russian)

E3S Web of Conferences **11**, 00008 (2016) DOI: 10.1051/e3sconf/20161100008
Solar-Terrestrial Relations and Physics of Earthquakes Precursors

First results of atmospheric electricity measurements by CS110 electric field meter at Paratunka observatory, Kamchatka

Sergey Khomutov[1,*], *Sergey* Smirnov[1], *Stanislav* Butin[1], and *Ilkham* Babakhanov[1]

[1] *Institute of Cosmophysical Researches and Radio Wave Propagation FEB RAS, Paratunka, Kamchatsky Kray, Russia*

Abstract. The results of continuous measurements of AEF potential gradient by CS110 fluxmeter at Paratunka observatory are presented.

1 Introduction

Atmospheric electric field (AEF) is the object of active scientific investigations. The experimental basis for such investigations is provided by monitoring measurements of AEF at observatories as well as special observations at temporary stations. However, at the present time, there are no standards for such measurements and international or national centers for data collection. Thus, reliability of the obtained results at the observatories is usually ensured by standard procedures of calibration and comparison of devices of different types.

At Paratunka Geophysical Observatory of IKIR FEB RAS, regular observations of AEF have been carried out since 1996 [1]. We use a widely spread sensor "Pole-2" (electric field meter, fluxmeter) to measure the electric field intensity [2]. It ensures reliable and qualitative results that is also due to the self-calibrations and occasional comparisons with other electric field meters, for example, with EF-4 or Gradient-3 [3]. However, this device is out of date instrumentation both structurally (element base and circuit technology) and physically in the result of degradation of the construction mechanical elements. Thereby, IKIR purchased a modern digital electric field meter CS110 (Campbell Scientific, Inc., USA) in November 2015 within the framework of Russian Scientific Foundation Grant. Besides the modernization of instrumentation for AEF monitoring, the task was to compare the results of measurements by electric field meter "Pole-2" with that carried out by the electric field meter CS110 which corresponds to the international standards in the area of AEF observations (for example [4]).

2 Instrumentation and operation conditions

The electric field meter CS110 is a device widely spread for AEF investigations, including lightning activity control. Measurements of the vertical gradient of AEF potential E_Z are carried out according to the same methodical principle as in "Pole-2" sensor: electric field modulation by a field mill. However, in comparison to "Pole-2" mill rotating with high speed, CS110 has a modulating mill with reciprocating motion that is provided by a stepper motor. Moreover, the measuring element of

*e-mail: khomutov@ikir.ru

 DOI: 10.1051/e3sconf/20161100008

Figure 1. Electric field meters at "Paratunka" observatory. Left panel: pillar with sensor "Pole-2" at its center (in front is a stadia rod to measure the snow depth), the technical hut with the electric field meter CS110 on the roof is in the background. Middle panel: the electric field meter CS110 on a mast on the roof of the technical hut, pillar with "Pole-2" sensor is in the background. Right panel: pillar and sensor "Pole-2" are covered with snow after a cyclone (March 23, 2016).

CS110 sensor is oriented downwards that protects its shutter from precipitation causing problems for "Pole-2" during strong snowfalls. CS110 sensitivity in the range of ±2200 V/m is 0.32 V/m, root mean square value is 0.42 V/m. There is a capability of automatic switch to an extended range of ±22300 V/m (with the loss of sensitivity of about one order). The electric field meter includes a logger CR1000 for control, data collection and exchange with external devices, for example, with a PC. Information transmission to a PC is realized via the WiFi. Inner timer synchronization is provided by PPS-signal of the build-in GPS-receiver. Leakage current, CS110 sensor temperature and humidity are being controlled. Buffer memory and battery supply ensure autonomous measurements within 7 days. Moreover, additional external sensors, for example meteorological ones (temperature, humidity and precipitation) can be connected to the logger.

The electric field meter CS110 (sensor, logger box, GPS-receiver and UPS), GPS antenna and the sensor of the outdoor temperature and humidity were mounted on a special steel pole with solid base on the roof of a technical hut (the area is about 6×6 m^2 with bituminous coating, Fig. 1). The sensor, the mast and the box with electronics are grounded. The electric field meter line supply is 220 V. The height of the measuring element of the sensor is 1.2 m above the roof, the height of the roof above the ground level is 3.7 m. The electric field meter "Pole-2" is installed about 19 m to the South from CS110 on a frame pier. The shutter height is about 3 m.

The recording laptop with LoggerNet software is placed inside the hut. The measurement frequency is 1 Hz. An output file includes time, potential gradient E_Z (V/m), measurement status, leakage current (nA), temperature and humidity inside the sensor (^{o}C, %), battery voltage of the build-in UPS. A separate file contains the data obtained by 10-minute averaging. Data from the logger buffer memory are recorded into files on schedule (hourly). LoggerNet software displays the results of current measurements in text and graphic forms in on-line mode.

3 Results

Regular observations by CS110 began in November 2015, i.e. they covered the winter with snowfalls and low temperatures which cause many problems in AEF measurements. Within the whole period

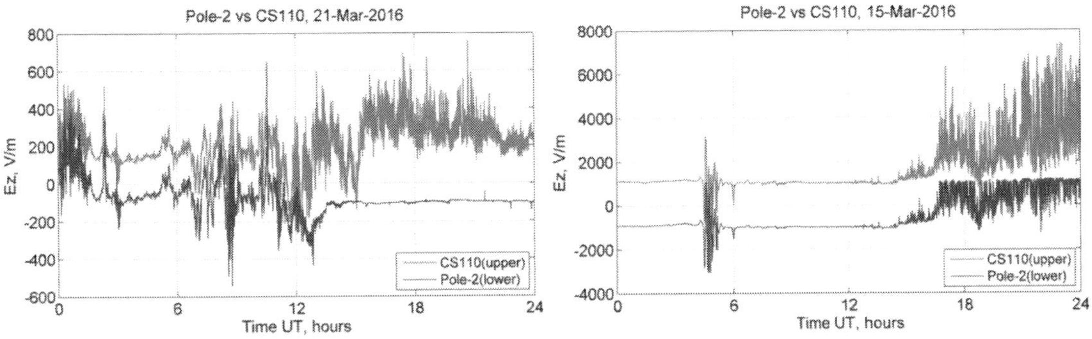

Figure 2. AEF records by electric field meters "Pole-2" and CS110 during extreme weather conditions. Left panel: during a heavy snowfall after 13 UT, "Pole-2" sensor was covered with snow. Right panel: light snow and storm wind after 16 UT cause displacement of "Pole-2" records outside the acceptable limit (CS110 is shifted for better presentation).

under consideration, there were no significant failures in instrumentation operation including the periods of power cutoffs. The initial calibration of the electric field meter CS110 was carried out by the manufacturer and is not required during ordinary operation. However, the results of measurements were not reduced to the ground level by the moment of preparation of the present paper since the corresponding reduction coefficient was not determined due to the snow.

As we noted before, cyclones during cold seasons are accompanied by heavy snowfalls when the snow depth may reach up to 0.5-1.0 m. Fig. 1 (right panel) shows "Pole-2" pier after a cyclone in March 2016. It is clear that the sensor is completely covered with snow. Its readings reduce to the zero. However, CS110 measurements were going on owing to the shutter oriented downwards (Fig. 2, left panel). The second advantage of measurements by CS110 is the extended range of measurements, to 22000 V/m, in comparison to "Pole-2" capabilities which range is limited to about ±2000 V/m (example of a record is in Fig. 2, right panel).

The results of measurements by two electric field meters during fair weather conditions are quite similar. As an example, Fig. 3 shows variations of E_Z for one day (February 4, 2016) and one hour (February 9, 2016).

As long as the coefficient of reduction to the ground level for CS110 was not determined, there are regular scale differences in the data which are eliminated from the curve in Fig. 3 by adjustment of the appropriate coefficient. It is clear from Fig. 3 that the results from the both devices are quite similar. For fast variations of E_Z, the difference variances reach 10-20% and they are likely to be caused by local AEF dynamics due to the spatial separation of the devices and different time of response on sudden disturbances. Amplitude-phase differences in the records on characteristic times of units-tens of seconds are quite often observed during light wind.

4 Conclusions

First regular measurements of potential gradient E_Z were carried out by a new electric field meter (fluxmeter) CS110 at "Paratunka" observatory, IKIR FEB RAS, during quite extreme weather conditions in winter-spring 2016 and without strict metrological basis. Nevertheless, the electric field meter proved to be a reliable and functional device having some advantages in comparison to the standard

 DOI: 10.1051/e3sconf/20161100008

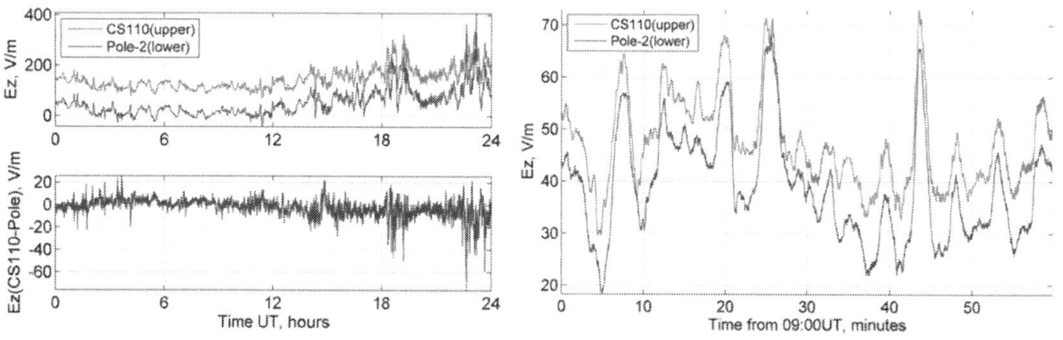

Figure 3. Comparison of variations of E_Z, obtained by the electric field meters "Pole-2" and CS110 during fair weather conditions. Left panel: daily record on February 4, 2016 (at the top) and their differences (at the bottom). Right panel: hour fragment of the record on February 9, 2016.

field mill "Pole-2" mainly during the observations when weather conditions are far from fair. The results of measurements by CS110 and "Pole-2" show good agreement during normal conditions. The observed differences may be explained by local features of AEF dynamics in the vicinity of the set of spatially separated devices.

In 2016 we plan to calibrate the CS110, taking into account the location, to correlate CS110 and "Pole-2", and to compare it with other electric field meters, with EF-4 for example.

This work was supported by the Russian Science Foundation Grant No. 14-11-00194.

References

[1] A.V. Buzevich, S.E. Smirnov, *23 General Assembly of the International Union of Geodesy and Geophysics* (2003), B81

[2] *Datchik napryazhennosti elektricheskogo polya "Pole-2". Tekhnicheskoe opisanie i instruktsiya po ekspluatatsii* (A.I. Voeykov GGO, Leningrad, 1991), 35

[3] R.R. Akbashev, P.P. Firstov, S.Yu. Khomutov, V.A. Efimov, *Problemy kompleksnogo geofizicheskogo monitoringa Dal'nego Vostoka Rossii* (GS RAN, Obninsk, 2015), 197–201 (in Russian)

[4] F.C. Magina et al., *XV International Conference on Atmospheric Electricity* (Norman, 2014), 1–11

The Shock Wave in the ionosphere during an Earthquake

Vladimir Kuznetsov
Institute of cosmophysical research and radio wave propagation FEB RAS 684034 Kamchatka, Russia

Abstract. Fundamentally new model of the shock wave (SW) generation in atmosphere and ionosphere during earthquake is proposed. The model proceeds from the idea of cooperative shock water crystallization in a cloud.

1 Introduction

Fundamentally new shock-wave model of an earthquake has being developed by the author during the last 25 years. Pr. Afraimovitch E.L. whom I presented my model in 2004 in IKIR approved it and acquainted me with his papers [2] on the shock acoustic waves (SAW) in atmosphere detected through the analysis of total electrons content (TEC) signals recorded by GPS-receivers (Fig.1).

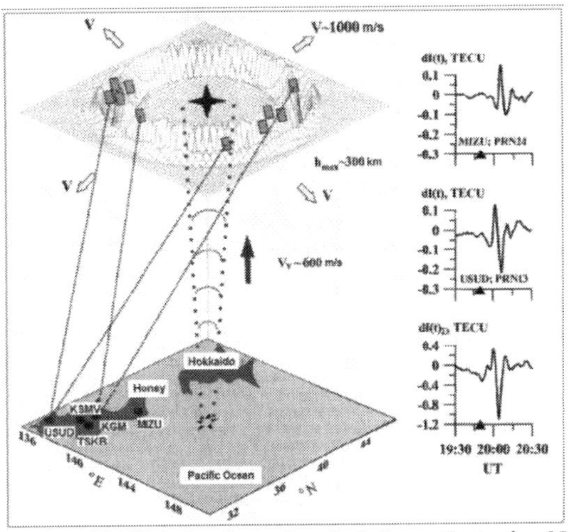

Fig. 1. Scheme of generation and detecting of SAW during the earthquake in Japan, September 25, 2003. Asterisk - earthquake epicenter, cross - secondary source, points - GPS-sensors, rectangles - ionospheric points. Right - TEC-variation during the earthquake near Hokkaido, September 25, 2003 [2].

Seeing the details of their research I didn't accept Pr. Afraimovitch' team interpretation of the phenomenon of SAW generation in ionosphere. What is known about the mechanism of ionospheric disturbances generation caused by seismic activity? As stated in [2, P. 338], such mechanism "is not clear". Modern approach to solving this problem consists in substituting the earthquake region for its point source [2, 3, 4]. The point underground source generates spherical elastic waves in mountain rock. The wave generation at the earth surface is similar to the generation of a strong shock when rock sections lift one after another and eventually return to their starting positions. The process is accompanied by the acoustic wave generation. The width of direction plot for the acoustic signal propagating from the perturbed Earth surface is very narrow [4]. Therefore, an acoustic disturbance penetrates up to the ionosphere altitudes within the narrow range of zenith angles. Moreover, the part of ionosphere above the epicenter is the source of ionospheric response. This model of ionosphere disturbances is developed in [3].

There are numerous results of observations of earthquake-induced strong ground motions and shock wave generation in various media. The fact that during powerful earthquakes the velocities of vertical ground motions are much less than the sound velocity in air opposes to the model of SAW generation proposed in [3]. The strong ground motions recorded during the Northridge earthquake will be taken as an example. Nevertheless, the observations on behavior of the electron density waves in ionosphere (Fig.1) prove the existence of SAW. This problem requires the development of principally new approaches.

Unfortunately, I began my research too late (2014) and I could not report about my results to Eduard Leontievitch who died (2009). I devote my manuscript to the blessed memory of Pr. Afraimovitch E.L.

Corresponding author: vvkuz38@mail.ru

2 Strong ground motions recorded during the Northridge earthquake

The Northridge earthquake (magnitude of 6.7, January 17, 1994) is one of the most studied. It occurred nearby Hollywood, the seismically dangerous and populous region of Los Angeles, with the large number of the seismic measuring equipment [5]. Data on three components of the strong ground motions obtained with the devices of Van Norman Complex [6] and the results on the motion magnitude and direction are especially essential.

The values of vertical and horizontal amplitudes and velocities of ground motions were obtained there. Amplitudes of horizontal motions were twice-three times more than vertical ones which were about $h \approx 10$ - 20 cm. The peak velocity of horizontal ground motion was about 177 cm/s, the vertical average velocity was 10 - 20 cm/s. The peak amplitude of horizontal motions was up to 40 cm. The hypothetical wavelength of vertical motions was $\lambda \approx$ not more than 50 cm. At the shock moment the ground has lifted on about ≈ 50 cm.

3 SAW record in ionosphere at the earthquake

Pr. Afraimovitch' team evaluated the height h of assumed source ($h \approx 300$ km). Not epicenter but the region of ionospheric layer at F_2 altitude over epicenter was the source of ionospheric disturbances. The secondary source initiates in 8-9 minutes after the main shock. The calculated velocity of signal propagation through atmosphere is about 600 m/s which is nearly twice as large as the experimental value for the speed of sound. It is easy to show that this model is incorrect.

4 Our model

Discussed model tries to interpret the occurrence of quantum entanglement between lithospheric and atmospheric processes. Entanglement may be activated with hydrogen bonds protons available in water and in many other geologic media such as kalicinite.

5 SW generation in lithosphere

Let's discuss the experiments on high-pressured kalicinite (KHCO3), the mineral in which hydrogen bonds show the capacity for quantum entanglement. On compression the structure of kalicinite hydrogen bonds changes simultaneously over the whole volume of the sample (Fig. 2) resulting in the change of its unit-cell volume (or density) and correspondingly (according to our model) in the SW generation. Similar data are obtained for other minerals containing hydrogen and oxygen (nitrogen, fluorine). By data of many authors such structural phase transition of hydrogen bonds happens instantly over the large volume and this phenomenon is called a cooperativity.

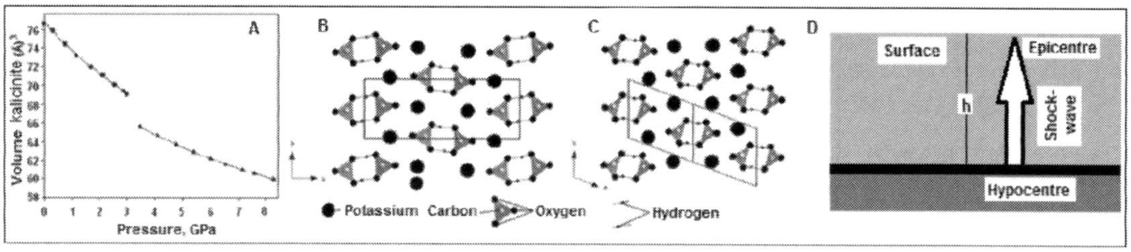

Fig. 2. The change of the unit-cell volume($\Delta V = V = 4\%$) for kalicinite at the pressure of 3.2 GPa - *A*, the change of kalicinite crystal structure - B,C. [7]. Shock wave generation inside the layer of lithosphere containing hydrogen bonds - D.

6 Hydrogen bond

A hydrogen bond occurs when the electron-lacked hydrogen covalently bound to the electronegative atom (oxygen, nitrogen, fluorine) interacts with lone-electron pair of another electronegative atom in the same molecule or in the other one. Cooperativity of hydrogen bonds in water in particular is responsible for the dependence of water features on the amount of hydrogen bonds. A sizeable quantity of hydrogen bonds in ice decreases with its melting up to their almost absence in boiling water. A lack of possibility to form hydrogen bonds would result in water transformation into vapor. Hydrogen bonding also plays an important role in lithosphere properties. Shifting of electron density at the line H…O (N, F, etc.) to the electronegative atom provides hydrogen bonding which produces interatomic distances O - O, O - N shorter than in hydrogen absence.

7 Earthquake energy

We are going to estimate the energy needed for kalicinite (KHCO3) compression (see Fig 2-A). The external pressure $p = 3$ GPa leads to a change in the sample volume $\Delta V/V \approx 5\%$ that is about 10^2 J/cm^3. In Fig. 2-B and C the so-called structural phase transition "order-disorder" is shown. In [7] the enthalpies of this transition are given for a wide range of crystals. We assume that energy of one broken hydrogen bond is about 0.5 eV (1 eV $\approx 10^{-19}$ J). Let's consider how the energy released at fracture (fracture, rotation, etc.) of hydrogen bonds is consistent with the estimate obtained from the relation $E = pV$. In minerals under the external pressure the presence of hydrogen bonds can provide the energy necessary for the shock wave generation.

8 Quantum entanglement

Quantum entanglement is a perspective research direction included into the list of 20 Big Scienti_c Ideas in "BBC Focus Special Edition, 2013". Here we try to link such notions as 'entanglement', 'decoherence', 'recoherence' and 'teleportation' with crystallization and different types of phase transitions that can occur in lithosphere and atmosphere.

Since we do not exploit measurement, we can capture 'entanglement' through collective decoherence without common environment. In our earthquake model the generation of the shock wave follows from decoherence of multiparticle quantum system and subsequent fast phase transitions. Moreover, decoherence can transform one quantum system into another which can have additional properties revealed in the form of an aftershock, teleportation of proton state and ect.

9 Quantum decoherence

Decoherence (D) of quantum system (QS) occurs every time when its state becomes entangled with a state of the environment. Thus, 'information' of QS is imprinted in a state of its environment. It is known that D is a physical process accompanied with the loss of coherence because of entanglement with the environment. Moreover, after decoherence QS gets classical features reflected into environment. Under natural conditions it is impossible to isolate multiparticle QS. Therefore, open quantum systems are common. Because of entanglement with the environment a coherent state transforms into a 'classical' mixed one. Moreover, an inverse transition called recoherence is possible. For recoherence there is only need in weakening of interaction between system and environment. Since there is 'information' about quantum states imprinted in the environment, we can observe transformations of 'classical' environment induced by decoherence or recoherence. Due to the no-go theorem it is impossible to statistically verify these transformations. This idea prohibits prediction of an earthquake.

10 Quantum teleportation

A quantum teleportation (QT) is a transition of quantum state for arbitrary long distances with the help of classical channel and separation of quantum pairs. We can a priori assume that there are entangled QS separated by a distance. Since we cannot exclude the classical interaction (classical channel), we deal with QT. In some cases the classical channel is not obvious, for example, it is not clear how protons in hydrogen bonds of supercooled water in Arctic cyclones are classically linked with protons in hydrogen-bonded camphora of a storm-glass in Novosibirsk [8].

Another example of QT is a 'connection' between the Sumatra earthquake in 2004 and typical fluctuation of atmospheric electric field in Italy [9]. There are numerous examples of 'superluminal connection' between far distanced separated earthquakes. QT can be the only reason of such connections. It seems that sprites, jerks and particles of high energy in water saturated atmosphere are generated because of QT. By the way, here a lighting bolt is a 'classical channel'.

11 Quantum nature of aftershocks

The earthquake time series (aftershocks, also) is flicker-noise (see the Gutenberg-Richter law). In recent years the study of open quantum systems showed that flicker-noise accompanies decoherence. It follows that the nature of earthquakes, including aftershocks, is quantum. If the conditions in a QS allows full decoherence, the entire system is converted into a classical one with generating shock wave. If there is only a partial decoherence, then the second, third and etc. SW may occur after formation of the first shock wave. It follows that the formation of the aftershock differs.

12 What is SAW?

Interesting video clips of model tests of a shock-acoustic "gun" are available online. SAW generated by this device leads to the destruction of a specially constructed models of buildings, located at a distance of several dozen meters from the cannon. In the chamber of the gun SAW is generated under the bulk detonation of explosives in the form of aerosols. Nothing is discarded from the "barrel" of the gun during the shot, the SW which effects the destruction is only emitted. These experiments can be considered as the mechanistic model of the considered natural phenomenon of generation of SAW. In our case, the shock crystallization of supercooled water in the cloud plays the role of a SAW generator (detonator).

13 Shock crystallization and generation of SAW

Our model of SAW is alternative to the conventional one, shown in Fig.3. The model is based on experimental and theoretical studies of the shock crystallization in supercooled water. The model simultaneously solves two problems: the occurrence of hole-punch clouds and the existence of "thunder in a clear sky" phenomenon. The energy difference between the pieces of ice and drops with ice nuclei (Fig. 3-b) determines the energy of SAW. In the cloud of supercooled water there are conditions for a shock crystallization (Fig. 3-a). QS of a cloud can be quantum entangled with hydrogen bonds in the lithosphere. Here SW in a cloud is an aftershock of the main earthquake.

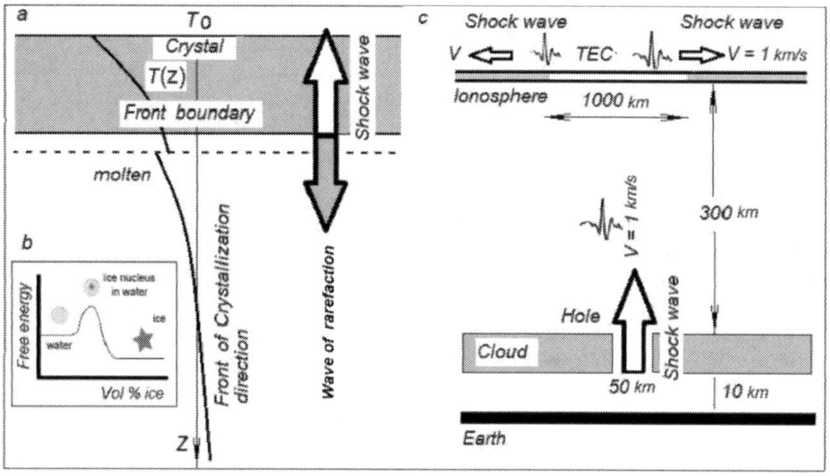

Fig. 3. The occurrence of crystallization front, rarefaction waves and SAW in supercooled water of clouds - a. Comparative particle energy (Free energy) of snowflakes, water droplets with the ice nuclei and iceless water droplets - b. Reaching ionosphere SAW is propagating through it. The hole in the cloud after the shock crystallization - c. Note that the formation of holes in atmosphere is a well studied phenomenon that is not explained until now.

14 Conclusion

The article considers one of the examples of natural phenomena, which Einstein once called "spooky action at a distance". In the Earth and solar-terrestrial physics almost all phenomena in varying degrees fall under this definition. However, in physics and in mathematics the approaches to solving problems in many-particle quantum entanglement and methods for their solution are successfully being developed.

References

1. V.V. Kuznetsov Physical Mesomec. **12** (in Russian) (2009)
2. E.L. Afraimovich, N.P. Perevalova GPS-monitoring of the Earth's upper atmosphere. Irkutsk: SC RRS SB RAMS. 480 p (in Russian) (2006)
3. R.V. Row, J. Geophys. Res. **72** (1967)
4. G.V Rudenko, A.M Uralov J. Atm. Terr. Phys. **57**. (1995)
5. J.P. Bardet, C. Davis, BSSA **86** (1996)
6. J.G. Anderson, G. Yu, BSSA. **86** (1996)
7. D.R. Allan, W.G. Marshall, C.R. Pulham American Mineralogist **92** (2007)
8. N.D. Kuznetsova, V.V. Kuznetsov J. of Science Forming Direction **3**. (in Russian) (2015)
9. H. Roder, T. Braun, W. Schuhmann, E.Boschi, R. Buttner, B. Zimanowski EOS. Trans. AGU. **86** (2005)

Peculiarities of geodeformation measurements of near surface sedimentary rocks

Igor Larionov[1,⋆] and *Yurii* Nepomnyashchiy[1,⋆⋆]

[1] *Institute of Cosmophysical Research and Radio Wave Propagation, Far Eastern Branch, Russian Academy of Sciences 684034, Kamchatskiy Kray, Paratunka, Mirnaya st., 7, Russia*

Abstract. The results of investigations of the deformation process in the near surface sedimentary rocks are presented. These investigations have been carried out in a seismically active region of Kamchatka peninsula since 2007. The peculiarity of the experiments is the application of a laser strainmeter-interferometer constructed according to the Michelson interferometer scheme.

1 Introduction

Seismotectonic process occurring as the result of accumulation and relaxation of lithospheric stress which is the consequence of motion and interaction of continental and oceanic plates is of great interest for investigation since it plays a significant role in many geophysical processes. To make diagnostics of deformations, we apply different methods of direct measurements by strainmeters. We also investigate such processes as acoustic emission generation, near ground electric field change, and register emissions of different gases.

2 Methods of investigation

One of the effective techniques for direct observations of deformations is the application of laser strainmeters. The operating principle of the simplest devices is that the change of strainmeter base causes additional phase increment in a wave of laser radiation. The method of measurement is as follows. Shift of interferometer mirrors arranged at the ends of a base l by the value $\lambda/2$ results in the change of interference pattern by one band, where λ is the light wave length on which the interferometer operates. Total relative shift equals $\Delta l = N(\lambda/2)$, where N is the number of bands of an interference pattern. The capabilities of the interference method are limited by the measurement accuracy of fraction shifts of a band ΔN which is determined by the parameter of interference pattern sharpness F_k and is characterized by the relation $F_k = \frac{\Delta\lambda}{\delta\lambda}$, that means that it is the relations of the distance between the maxima and maximum half width $\delta\lambda$.

The advantage of laser strainmeters in comparison to mechanic analogies is the absence of a mechanical sensing element. Effect of meteorological parameter variations on the device is reduced to

⋆e-mail: igor@ikir.ru
⋆⋆e-mail: vicekam@gmail.com

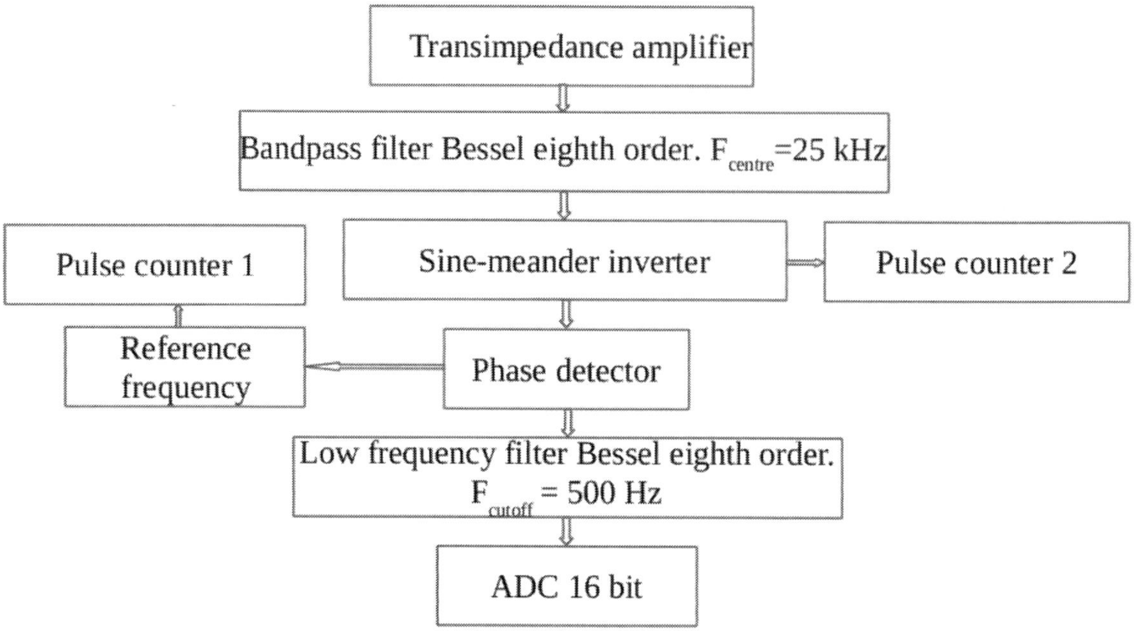

Figure 1. Block diagram of straimeter registration system.

their impact on the change of a laser beam optical path. When applying sealed vacuum-processed light guides, the accuracy for the measurement of relative deformations of the Earth's crust is 10^{-10} – 10^{-11} for best interferometers. Accuracy constraints determined by the effect of weather parameter variations are applied to the measurements carried out by "open" strainmeters without light guides. Wind, precipitation, air temperature and pressure have the strongest effect [1]. To exclude the wind and precipitation effect, we use special cover. Air pressure and temperature changes are quite slow processes which occur within several hours which allows us to make estimates of more fast deformation processes. Based on the calculated data, a strainmeter installed in such conditions has the accuracy of relative deformation measurements not less than 10^{-8}. As the observation results show, during the deformations of such order and more, the effects in sedimentary rocks generate acoustic signals in the frequency range from tens of hertz to the first ten of kilohertz [2–4].

Since 2007, the Laboratory of Acoustic Research of the Institute of Cosmpphysical Rasearch and Radio Wave Propagation (IKIR) FEB RAS has carried out geodynamic observations of the near surface sedimentary rocks as they are characterized by weak strength, high plasticity and are the most available to investigate deformation processes [3]. A laser strainmeter-interferometer of unequal-type constructed according to Michelson interferometer scheme developed at TOI FEB RAS was used to register deformations until 2013 [2, 4].

Due to the breakage of some key elements, a new strainmeter-interferometer was developed at the Laboratory of Acoustic Research of IKIR. It was constructed according to the same optical scheme and was put into operation in 2016. The block diagram of the strainmeter registration system is shown in Fig. 1.

The electric signal corresponding to the change of light intensity in the interference pattern is formed on a photodiode which is thereupon transformed into voltage and amplified on a transimpedance amplifier. A frequency range carrying the useful information is separated from the spectrum of the obtained electric signal by a band filter of the 8-th order. The pick of a pass band corresponds to the modulation frequency of 25 kHz. Then a signal is transformed into a square signal as long as the amplitude in this case does not carry any useful information. After the transformation, the signal comes to a pulse counter 2 and on input 2 of a phase detector. The reference signal modulating the phase shift between the reference and the measuring laser beams comes to the pulse counter 1 and on input 1 of the phase detector. The phase detector generates a signal proportional to the phase difference within 2π between the reference and the interference signals. The number of phase shift transitions across 2π is determined by the difference of readings of pulse counters 1 and 2. To digitize a signal formed by the phase detector, we use a 16-bit ADC together with LF filter of the 8-th order with the cut-off frequency of 500 Hz. The ADC sampling rate is 1 kHz.

In the result of digitization, two files are formed in a PC. One of them contains a signal from the phase detector. The other contains the number and the direction of phase transition across 2π. Then, taking into account the data of phase transitions, a processing program removes them from the initial signal of phase detector and forms daily files of absolute deformation changes.

It is generally accepted in data interpretation to apply dimensionless quantities of relative deformations that is why the data on graphs are shown relatively a strainmeter gauge length. In the analysis of weather parameter effects, we used the data from a temperature sensor fixed directly on the basis of the strainmeter optical system and from a weather station located at the distance of 20 m from the strainmeter. Such parameters of the environment as air temperature, humidity, and pressure change slowly during a day and do not affect fast deformation processes. Wind velocity makes significant effects on the strainmeter readings which appear as increases of signal amplitude oscillations.

3 Results of the investigation

The main direction in the investigation of near surface sedimentary rock deformations is the relation of local deformation process and its derivatives, such as generation of acoustic emission, near ground electric field changes and so on. In the previous papers of the author [2–5] such a relation was confirmed. Thus, in the construction of the strainmeter, a simple scheme of strainmeter-interferometer of the first generation with a frequency-stabilized laser was chosen.

Fig. 2 and 3 show examples of registration of seismic events of different energy classes which occurred at about the same distance from "Karymshina" observation site. To determine the earthquakes characteristics, the data of Kamchatka Branch of the Geophysical Service RAS were applied.

The strainmeter data are shown with the recording frequency of 1 kHz. The second channel of registration is used to display the data. Geoacoustic emission signals are registered with the frequency of 44 kHz. Fig. 2 and 3 illustrate only its low-frequency part. It is clear from the figures that the most powerful earthquake has the strongest signal at both measurement systems. These data allow us to estimate the correctness of adjustment and operation of the strainmeter registration system.

In the case of the earthquake on March 14, 2016, an increase in the deformation process rate was observed. A hydrophone installed in the intermediate vicinity from a reflector of the strainmeter measuring arm registered the increases of the amplitude and the number of acoustic pulses in the frequency range of 70-200 Hz (Fig.4). No wind, precipitation, sudden changes of air pressure and temperature were registered during this period. It allows us to consider this case as a local process of stress accumulation in rocks before the earthquake.

 DOI: 10.1051/e3sconf/20161100010

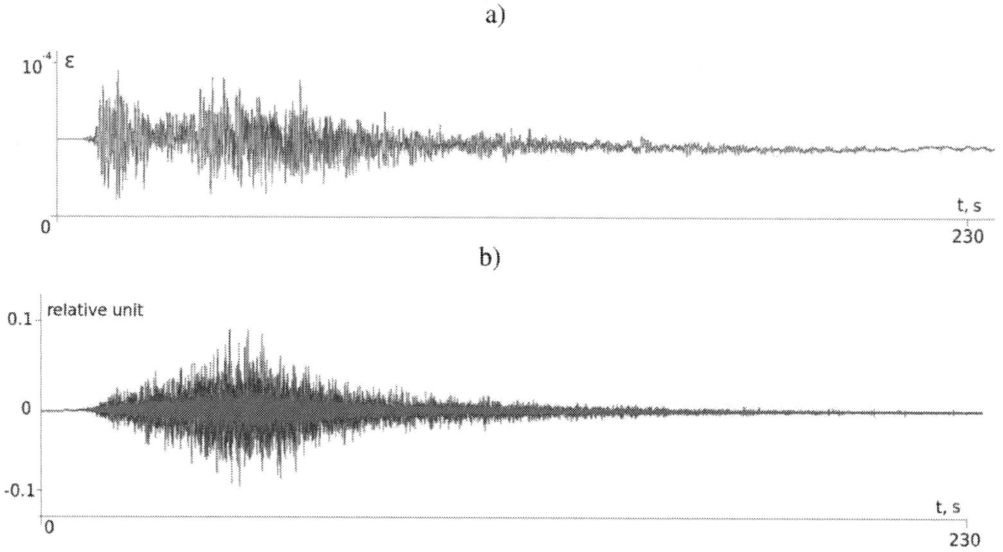

Figure 2. Seismic event on January 30, 2016 at 03:25 UT, energy class is 15,7, distance to the station is 129 km. a) Strainmeter, b) Geoacoustic emission.

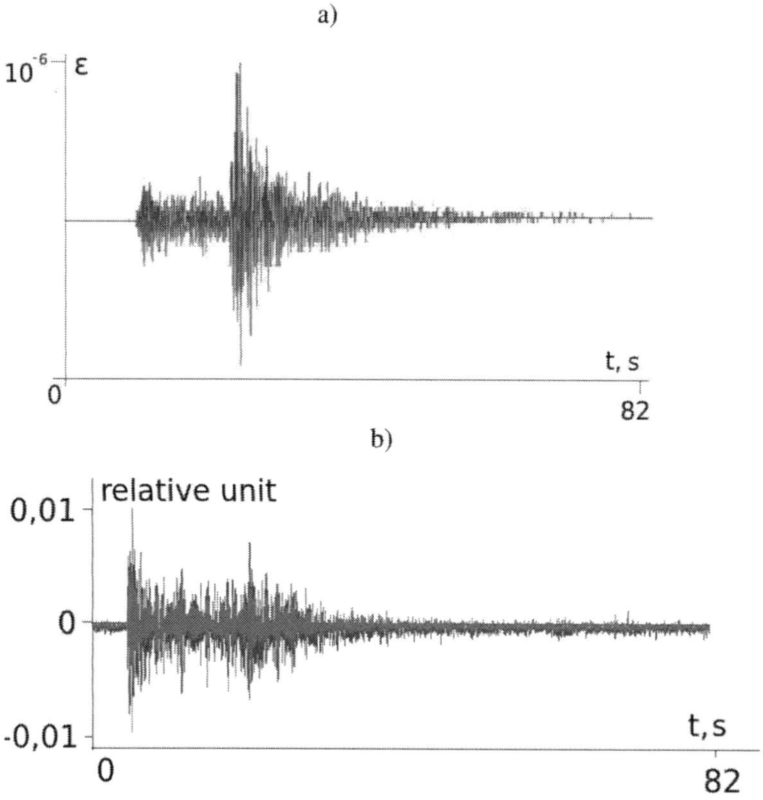

Figure 3. Seismic event on March 14, 2016 at 21:50 UT, energy class is 11,2, distance to the station is 133 km. a) Strainmeter, b) Geoacoustic emission.

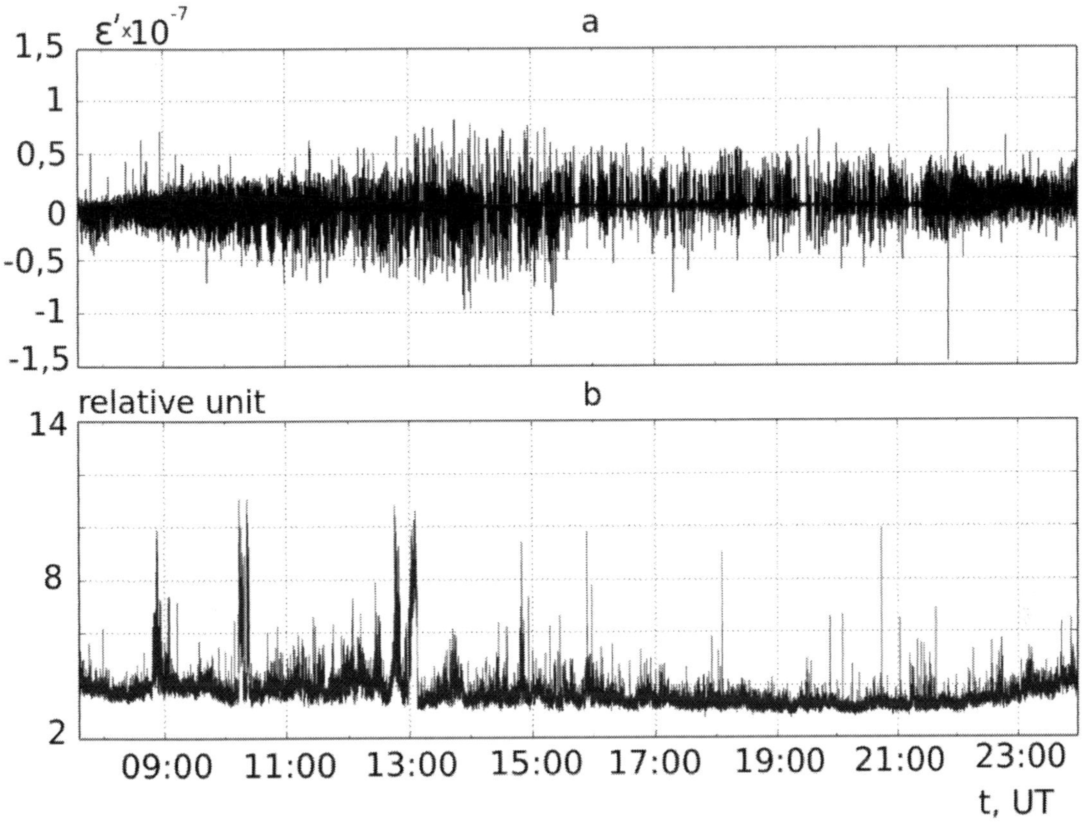

Figure 4. Period from 7:30 to 24:00 UT on March 14, 2016. a) Deformation rate, b) Geoacoustic emission in the frequency range of 70-200 Hz.

4 Conclusions

Thus, it has been shown that deformation measurements at "Karymshina" observation site have been resumed to their full extent which allowed us to investigate deformations at different stages of seismic activity during fair weather conditions.

References

[1] S. G. Dolgikh, et al., Gornotaezhnoe geophysical system, Instruments and Experimental Techniques, **55**, 298-299 (2012)

[2] G. I. Dolgikh, V. A. Shvets, V. A. Chupin, S. V. Yakovenko, A. V. Kuptsov, I. A. Larionov, Y. V. Marapulets, B. M. Shevtsov, O. P Shirokov, Deformation and acoustic precursors of earthquakes, Dokl. Earth Sci. **413** 281–285 (2007)

[3] I. A. Larionov, Y. V. Marapulets, B. M. Shevtsov, Features of the Earth surface deformations in the Kamchatka peninsula and their relation to geoacoustic emission, Solid Earth, http://dx.doi.org/10.5194/se-5-1293-2014, **5**, 1293–1300 (2014)

[4] Yu. V. Marapulets, B. M. Shevtsov, I. A. Larionov, M. A. Mishchenko, A. O. Shcherbina, A. A. Solodchuk, Geoacoustic emission response to deformation processes activation during earthquake preparation, Russian Journal of Pacific Geology,**6**, 457-464 (2012)

[5] Yu. V. Marapulets, O. P. Rulenko, I. A. Larionov, M. A. Mishchenko, Odnovremennyy otklik vysokochastotnoy geoakusticheskoy emissii i atmosfernogo elektricheskogo polya na deformirovanie pripoverkhnostnykh osadochnykh porod, DAN, T.440, **3**, (2011),403-406

Detection of the whistlers in the VLF-spectrograms

Ekaterina Malysh[1,*] and *Nina* Cherneva[2]

[1] *Vitus Bering Kamchatka State University, 683031, Petropavlovsk-Kamchatsky, Russia*
[2] *Institute of Cosmophysical Researches and Radio Wave Propagation FEB RAS, Paratunka, Kamchatsky Kray, Russia*

Abstract. The whistlers pattern recognition algorithm in the spectrogram of VLF-radiation and algorithm of the filtering of pulse signals in the spectrograms are proposed.

1 Introduction

The whistlers it is a large part of the natural electromagnetic emission. They are formed as a result of frequency dispersion of the lightning pulse in the propagation of this pulse in the magnetospheric plasma. Whistlers are manifested in the spectrograms of VLF-signals as a structures (patterns) which having a characteristic curving shape [1]. The form of the dispersion curve is determined by the plasma parameters, so whistlers are natural markers of status Earth's plasmasphere [2], and their monitoring is a promising area of research of space weather.

The monitoring of whistlers on a planetary scale is carried stations of AWDANet, which are coordinated within the framework of the international project "Plasmon" [3]. One of the receivers of the AWDANet operates in site "Karymshina" of the Institute of Cosmophysics Research and Radio Wave Propagation (IKIR) FEB RAS.

The monitoring of whistlers in the IKIR also is conducted by means of the complex of its own design, which implements the automatic whistlers recognition system [4]. However, this system does not allow the detection of whistlers in real-time mode, which requires the development of new recognition systems.

In this paper, we describe an algorithm for image recognition of the whistlers in the spectrograms. Initial data for the problem of recognition in monitoring system are 15-minute wav-file of the VLF-signal. Viewed algorithm processes a 4-seconds segment of the file. The main idea of the algorithm is based on recognition "rectified" images and was described earlier in [5].

2 Median filtering of the spectrograms

Let us consider this segment of the spectrogram as a two-dimensional array $\mathbf{A} = \{A_{ij}\}$, $i = 0, \ldots, N-1$, $j = 0, \ldots, M-1$, where $N = 513$ and $M = 172$. The index i enumerates discrete frequencies f_i, and index j enumerates discretes t_j. An example of such a spectrogram containing one of whistler, shown in Fig. 1 top.

*e-mail: roswell-47@mail.ru

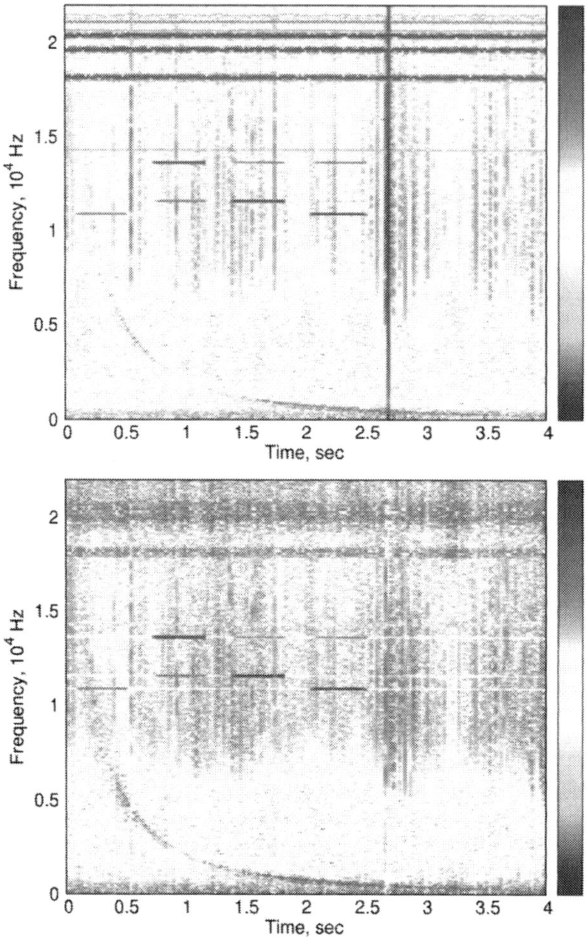

Figure 1. The spectrogram of VLF signal segment containing whistler: initial (top panel) and after median filtering (bottom panel).

The problem of whistler detection may be regarded as a problem of pattern recognition in a two-dimensional image. The desired pattern is a curve with a characteristic form defined in the plane (t, f) by the ratio $t - t_0 \sim f^{-1/2}$ [1].

From Fig. 1 can clearly see that in addition to whistler in the analyzed signal there are δ-like components (lightning pulses, so-called atmospherics) and narrow-band components. And those and others are noise for the problem of recognition. They are appear in the spectrogram in the form of lengthy linear structures, vertical and horizontal, respectively.

There is a need, above all, to quell this noise structures, and select a statistically significant points of the histogram.

It is well known, that the effective tool for cancellation lengthy noise in the images is the median filtering [6]. The median filters belongs to a class of nonlinear filters and keeps the sharpness of the image, as they are based on the order statistics, which are robust to strong emissions in data. The paper [7] was offered a modified median filtering algorithm, which allows to suppress lengthy vertical noise structures in satellite images. This algorithm has been adapted by us to solve the problem of cancellation lengthy vertical and horizontal structures to the spectrograms.

First, we make two one-dimensional arrays **X** and **Y**, whose elements are the average values of the original spectrogram array in rows and columns, respectively:

$$X_i = \frac{1}{N}\sum_{k=0}^{M-1} A_{ik}, \quad Y_j = \frac{1}{M}\sum_{k=0}^{N-1} A_{kj}. \tag{1}$$

As a result, the each vertical noise structure is concentrated in several neighboring elements of array **X**, and the each horizontal - in several neighboring elements array **Y**. The strong emissions in these arrays will correspond to filterable noise structures.

Then, each of these arrays is subjected to one-dimensional median filtering:

$$\begin{aligned}
&\hat{X}_i = \text{median}\{X_{i+k}| \quad k = -L_X, \ldots, L_X\}, \quad i = L_X, \ldots, N - L_X - 1,\\
&\hat{X}_i = \hat{X}_{L_X}, \quad i = 0, \ldots, L_X - 1,\\
&\hat{X}_i = \hat{X}_{N-L_X-1}, \quad i = N - L_X, \ldots, N - 1,\\
&\hat{Y}_j = \text{median}\{Y_{j+k}| \quad k = -L_Y, \ldots, L_Y\}, \quad j = L_Y, \ldots, M - L_Y - 1,\\
&\hat{Y}_j = \hat{Y}_{L_Y}, \quad j = 0, \ldots, L_Y - 1,\\
&\hat{Y}_j = \hat{Y}_{M-L_Y-1}, \quad j = M - L_Y, \ldots, M - 1,
\end{aligned} \tag{2}$$

where $2L_X + 1$ and $2L_Y + 1$ are the lengths of filter masks. Then arrays noise is calculated $\Delta\mathbf{X} = \mathbf{X} - \hat{\mathbf{X}}$ and $\Delta\mathbf{Y} = \mathbf{Y} - \hat{\mathbf{Y}}$.

Filter spectrogram $\hat{\mathbf{A}}$ define as $\hat{A}_{ij} = A_{ij} - (\Delta X_i + \Delta Y_j)/2$. This spectrogram is shown in Fig. 1 bottom. It can be seen that the proposed algorithm of median filtering images really effectively suppresses atmospherics and narrowband noise, and the pattern of whistler remains clear. The short radio signals (~ 0.5 sec.) also remained.

3 Selection of significant points of the spectrogram

On next step, we selected of significant points for each column in the filtered spectrogram. We consider the squares of elements of the column as the Schuster periodogram corresponding segment of the VLF signal, and checks the hypothesis that this segment is white noise on the significance level α [8].

To do this, we calculate for each j-th column of the spectrogram $\hat{\mathbf{A}}$ variance σ_j^2 through its spectral representation:

$$\sigma_j^2 = 2\sum_{i=0}^{N-1} \hat{A}_{ij}^2. \tag{3}$$

Furthermore, if $\hat{A}_{ij}^2 < -\sigma_j^2 (\ln\alpha)/N$, then the point $\hat{A}_{ij}$ is considered noise, otherwise case – significant.

Of the pairs of indices (i, j) of significant elements we formed the dynamic one-dimensional array **P** with elements $P_m = (i_m, j_m)$, whose length is equal to the number of significant elements of the spectrogram.

This is done to acceleration of processing. The thing is further recognition algorithm is working with arrays of $\mathbf{P} = \{P_m\}$, $\mathbf{t} = \{t_j\}$ and $\mathbf{f} = \{f_i\}$, but not with points of the spectrogram. The element P_m corresponds to the point $\hat{A}_{i_m j_m}$ on the filtered spectrogram. Next we are identifying P_m and $\hat{A}_{i_m j_m}$.

The transition to these arrays provides a significant reduction in the volume of data being processed further. Analysis of typical spectrograms showed that the length of the array $\mathbf{P}$ is approximately by an order less than that of the initial spectrogram.

On top panel of the Fig. 2 by points shows the elements selected for the spectrogram $\hat{\mathbf{A}}$. It can be seen that the image of whistler is well looked through after selection.

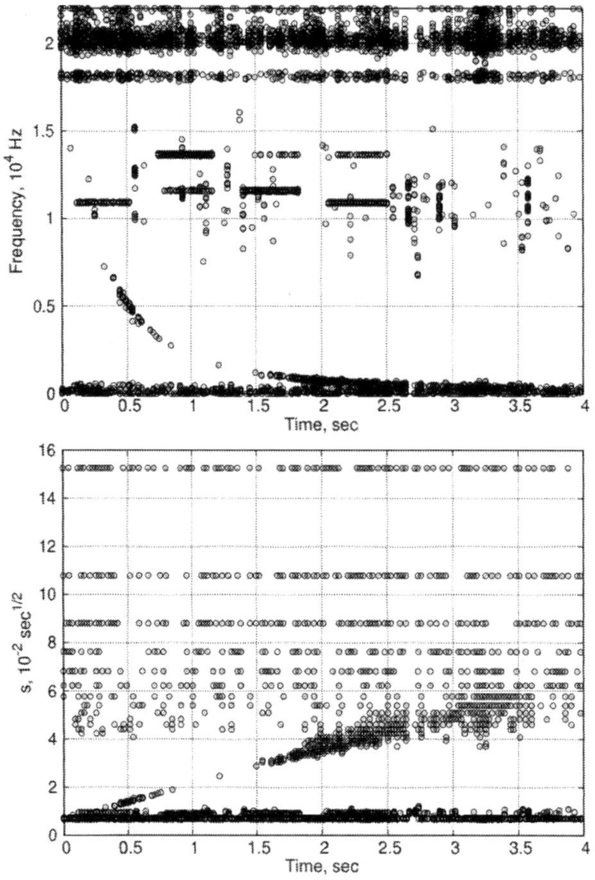

Figure 2. Significant points of the spectrogram in the plane (t, f) (top panel) and in the plane (t, s) (bottom panel).

4 Recognition algorithm

From the theory of VLF waves propagation in the magnetospheric plasma is known [1], that the pattern of whistler, has come at the time τ, is well described by the equation

$$t - \tau = D / \sqrt{f}, \tag{4}$$

where the coefficient $D > 0$ is called dispersion. From (4) it is clear that, if we introduce the transformation $s = f^{-1/2}$, then in the plane (t, s) pattern of the whistler is rectified, and Eqs. (4) corresponds to the equation in this plane $t - Ds - \tau = 0$, i.e. equation inclined straight, which forming a non-zero acute angle with the positive direction of the time axis. The points $Q_m = \left(t_{j_m}, s_{i_m}\right)$ corresponding to the elements of the $P_m = (i_m, j_m)$ are shown in bottom panel of Fig. 2.

The straight line recognition problem is algorithmically simpler, than the recognition of the curve. Therefore, more convenient to detected whistler in the plane (t, s). Then, an indication of the presence of the whistler (whistlers) is a grouping of points near the straight line (lines) in this plane. And we are talking only about the slanted lines, as the remains of the narrow-band noise and radio signals after median filtering and selection of significant elements continue to manifest themselves in the form of horizontal rectilinear structures.

Of the many variants of the equations of this direct lines, is best suited for our purposes

$$t \sin\varphi - s\cos\varphi - d = 0, \tag{5}$$

where φ - angle formed by a straight line with the axis of time, and d - generalized distance from the line to the origin. The distance it takes a positive, if the line crosses the s-axis in a positive point, and negative otherwise.

Processing a large number of spectrograms showed that the real value $\varphi \in \left[1^0; 10^0\right]$. It is also clear that $d \geq d_{\min} = -\max\{t_{j_m}\}$ and $d \leq d_{\max} = \max\{s_{i_m}\}$.

A pencil of lines passing through the point Q_M, defined by the equation

$$t_{j_m} \sin\varphi - s_{i_m} \cos\varphi = d, \tag{6}$$

describing the line in the area $1^0 < \varphi < 10^0$, $d_{\min} \leq d \leq d_{\max}$ in the plane (φ, d). Then the straight line (whistler pattern), passing through several points of Q_m, will correspond to the point of intersection of these lines in (φ, d). Therefore, an indication of the proximity of the points to a one straight line in the plane (t, s) is the existence of a small area in the plane (φ, d), that is crossed these a line. Moreover, smaller the area, the closer to straight line arranged the point in the (t, s), and inversely.

We introduce a analytical grid $\{\varphi_k\}$ and $\{d_l\}$ on the set of possible values of angles and distances. For each point Q_m and every angle φ_k from the equation (6) define the distance d and rounded up to the nearest d_l.

The result is a two-dimensional histogram of the pairs (φ, d) distribution for the points Q_m. If the value histogram for the pair (φ_k, d_l) is large, then a lot of Q_m is near the corresponding line. Thus, each significant maximum of the histogram can be associated with whistler.

Selection of significant maximums may also be performed as described above on the basis of the Schuster theorem.

5 Conclusion

In this paper we describe an algorithm for whistlers pattern recognition in the spectrogram of VLF radiation, consisting of four stages: median filtering; selection of significant points; coordinate transformation with a view to "rectifying" the whistler pattern; recognition "rectified" whistler.

The recognition algorithm testing showed that processing of 225 spectrograms for 15-minute wav-file requires less than 4 minutes, which ensures significant reserve of processing time. Now we develop a neural network to detect whistlers in automatic mode on the basis of the described algorithm.

References

[1] B. N. Gershman, V. A. Ugarov, Sov. Phys. Usp., **3**, A07222 (1961)

[2] J. Lichtenberger, Jour. Geophys. Res.: Space Physics, **114**, A07222 (2009)

[3] http://plasmon.elte.hu/

[4] N. V. Cherneva, V. P. Sivokon', I. V. Agranat, *Problems of Geocosmos. Proc. of the 9th Int. Conf.*, 102–106 (2012)

[5] E. A. Malysh, Bulletin KRASEC. Phys.&Math. Sci., **11:2**, 82–87 (2015)

[6] R. C. Gonzales, R. E. Woods, *Digital Image Processing* (Prentice-Hall Inc., Upper Saddle River, 2002), 782.

[7] Yu. M. Gektin, A. A. Zaytsev, Trudy MFTI, **6:4**, 103–106 (2014)

[8] S. L. Marpl-jr., *Digital Spectral Analysis: With Applications* (Prentice-Hall Inc., New Jersey, 1987), 523

E3S Web of Conferences **11**, 00012 (2016) DOI: 10.1051/e3sconf/20161100012
Solar-Terrestrial Relations and Physics of Earthquakes Precursors

Analysis of ionospheric parameters during Solar events and geomagnetic storms

Oksana Mandrikova[1,*], *Yury* Polozov[1], *Igor* Solovev[1], and *Nadezhda* Fetisova[1]

[1] *Institute of Cosmophysical Researches and Radio Wave Propagation FEB RAS, Paratunka, Kamchatsky Kray, Russia*

Abstract. The paper shows new methods of analysis of ionospheric and magnetic data applying the models of multicomponent constructions (MCM models) developed by the authors. Based on ground station data, the analysis of ionospheric and magnetic data during increased solar activity was carried out.

1 Introduction

During increased solar activity, anomalous changes characterizing the processes in the Earth magnetosphere and ionosphere may be observed in the recorded ionospheric parameters [1–3]. The anomalous changes manifest themselves as significant deviations (increase or decrease) of electron concentration in relation to the characteristic level. The times of occurrence, relative values and duration of ionospheric anomalies depend on many factors. In seismically active regions of the Earth, ionospheric anomalies may also be determined by increased seismic activity [1, 4]. In spite of the active development of technologies for monitoring of the near-Earth space and methods for data analysis, capabilities of estimation and forecast of ionosphere state are quite limited at the present time. Thus, development of empirical methods and technical means for their detection is of great scientific and applied significance [1, 5].

The problems of analysis of ionosphere state and detection of anomalies are solved by many authors [2, 3, 5–8]. Among the general approaches we can emphasize the traditional moving median method [1, 9], empirical models of the ionosphere [2, 3, 5–7], application of neural networks [2, 8, 10], and wavelet transform [10–14]. The most developed empirical model of the ionosphere is the International Reference Ionosphere (IRI) model [5], which is based on a wide range of ground and space data. The IRI accuracy for estimation of the parameters in a certain region much depends on data availability and may give results with large deviations from experimental data [5]. Thus, a more accurate forecast based on the IRI model is obtained for midlatitueds in comparison to equatorial and auroral latitudes. It was shown in the papers [2, 7] that the IRI model accuracy much depends on the level of solar activity and goes down as the latter grows. Compared to the IRI model, the new developments of empirical models applying neural networks [2, 8] allow us to improve significantly the quality of forecast. They are easily realized in automatic mode and are quite flexible. However, these models refer to the class of "black box" models, so to describe an attribute space they require

*e-mail: oksanam1@mail.ru

long learning samples prone to overfitting and can show unexpected results in case of very noisy data. In the paper, we use a complex approach based on wavelet transform methods and their combinations with classical autoregressive approaches and neural networks. In the previous papers [4, 12–14] we showed that application of classical autoregressive methods [15] in combination with modern methods for pattern recognition allow us to obtain quite exact estimates and they are easily realized in automatic mode. The main advantage of the suggested approach is the mathematical validity and, as a consequence, the possibility to receive the results with defined confidence probability.

Ionosphere parameters were analyzed together with geomagnetic field data (www.ikir.ru). Methods described in the paper [16] were used for processing. The obtained results were compared with solar wind parameters (http://www.srl.caltech.edu/ACE/ASC).

2 Description of methods

1. MCM model-1 of ionospheric parameter time series based on wavelets and autoregressive models.

Step 1. On the basis of multiple-scale decompositions (MSD), the time series f_o is presented in the form of a linear combination of different-scale components: smoothed one $f[2^{-m}t]$ of the scale m and detailing ones $g[2^j t]$ of the scales $j = \overline{-1, -m}$ [4]:

$$f_o(t) = \sum_{j=-1}^{-m} g[2^j t] + f[2^{-m}t], \tag{1}$$

where $g[2^j t] = \sum_n d_{j,n}\Psi_{j,n}(t)$, $d_{j,n} = \langle f, \Psi_{j,n}\rangle$, Ψ is the basic wavelet, j is resolution, $f[2^{-m}t] = \sum_k c_{-m,n}\phi_{-m,n}(t)$, $c_{-m,n} = \langle f, \phi_{-m,n}\rangle$, ϕ is the smoothing scaling function. The inferior index 0 corresponds to the data initial resolution.

Step 2. To obtain the components of the series $f[2^{-m}t]$, $m = \overline{1, M}$, we identify autoregressive models (the procedure of identification is described in detail in the papers [12, 14]).

Step 3. We join the obtained models into a multicomponent construction (MCM model-1) describing data time variation:

$$f_o(t) = \sum_{\mu=\overline{1,T}} \sum_{k=\overline{1,N^{\mu}_j}} s^{\mu}_{j,k} b^{\mu}_{j,k}(t), \tag{2}$$

where $s^{\mu}_{j,k} = \sum_{l=1}^{p^{\mu}_j} \gamma^{\mu}_{j,l}\omega^{\mu}_{j,k-l} - \sum_{n=1}^{h^{\mu}_j} \theta^{\mu}_{j,n} a^{\mu}_{j,k-n}$ is the assessed value of μ component; p^{μ}_j, $\gamma^{\mu}_{j,l}$ are order and parameters of autoregression of μ component; h^{μ}_j, $\theta^{\mu}_{j,k}$ are order and parameters of the moving average of μ component; $\omega^{\mu}_{j,k} = \nabla^{\nu^{\mu}}\beta^{\mu}_{j,k}$; ν^{μ} is the difference order of μ component; $\beta^{1}_{j,k} = c_{j,k}$, $\beta^{\mu}_{j,k} = d_{j,k}$, $\mu = \overline{2, T}$; T is the number of the modulated components; $a^{\mu}_{j,k}$ are residual errors of the model of μ component; N^{μ}_j is the length of μ component; $b^{1}_{j,k} = \phi_{j,k}$ is the scaling function; $b^{\mu}_{j,k} = \Psi_{j,k}$, $\mu = \overline{2, T}$ is the wavelet basis of μ component.

During the anomalous behavior of ionospheric parameters, their structure will change and, as a consequence, errors of MCM models will increase. Thus, the procedure of anomaly detection is based on the following conditional test:

$$\varepsilon_{\mu} = \sum_{q=1}^{Q_{\mu}} |a^{\mu}_{j,k+q}| > T_{\mu}, \tag{3}$$

where Q_{μ} is the length of prediction of the data based on the model of μ component; $a^{\mu}_{j,k+q} = s^{\mu,fact}_{j,k+q} - s^{\mu,forecast}_{j,k+q}$ are the residual errors of the model of μ component at the point $k+q$, T_{μ} is a threshold value of μ component determining the presence of an anomaly.

2. MCM model-2 of ionospheric parameter time series base on wavelets and neural networks

Step 1. On the bases of MSD (see presentation (1)), we obtain a smoothed component $f[2^{-m}t]$ of the series f_OF2. To exclude the tip effect, the initial set of data is extended by mirror segments (on the basis of function mirror reflection) with the length of $2^{-m} * q$ of counts (the size of a neighborhood containing the tip effect on a scale level $-m$ is determined by the formula $h_{-j} = 2^{-m} * q$, where q is the size of the applied wavelet).

Step 2. We perform wavelet reconstruction of the component $f[2^{-m}t]$ to the initial resolution removing mirror segments and obtain its presentation in the form: $f_0(t) = \sum_k c_{0,k}\phi_{0,k}(t)$.

Step 3. The obtained vector of the coefficients is $\{c_{0,k}\}_{k=1}^{K}$, where K is the vector length which we divide into blocks: $\{c_{0,k}\}_{k=1}^{S}, \{c_{0,k}\}_{k=2}^{S+1}, ..., \{c_{0,k}\}_{k=K-S}^{K}$. The length of a block is $S = 24$ (determined according to the diurnal variations of hourly data of f_OF2). Applying the blocks, we form a neural network (the procedure of construction of a neural network is described in the paper [17]). The quality criteria of network learning is: $E_A = \left(\frac{1}{R}\sqrt{\sum_{r=1}^{R} e_{A,r}^2(l)}\right) < \varepsilon_A$, where $\varepsilon_A > 0$ is a prescribed value, $e_{A,r}(l) = \hat{c}_{0,l}^r - c_{0,l}^r$ is the solution error at discrete instant of time l with the time of prediction r, $c_{0,l}^r$ is the desired output value and $\hat{c}_{0,l}^r$ is the real output value of the network, R is the length of the output vector of the network. Application of a neural network allows us to reproduce characteristic variations of the series f_OF2. During anomalous changes, neural network errors will be increased and they may be detected by the following conditional test: $E_A = T$, where T is the threshold value determining the presence of an anomaly. During the work, MCM models were developed separately for each registration site. Taking into account the seasonal changes of f_OF2 variation, data for different seasons were modulated separately. As long as the ionosphere state is reflected on geomagnetic field, we applied ionospheric data for time intervals of relatively calm geomagnetic field (intervals within which the total one-day K-index did not exceed 18) to model f_OF2 typical variations. **To analyze ionospheric parameters in detail**, we applied estimation solutions based on continuous wavelet transform [18]

$$(W_\Psi f)(b,a) := |a|^{-1/2}\int_{-\infty}^{\infty} f(t)\Psi\left(\frac{t-b}{a}\right)dt,$$

Ψ is the wavelet, $f \in L^2(R)$, $a, b \in R$, $a \neq 0$.

If we consider the series f_OF2 as f, detection of ionospheric anomalies may be based on the application of the following decision rule [12]:

$$P_{T_a}(W_\Psi f_{b,a}) = \begin{cases} W_\Psi f_{b,a}, & \text{if} \quad |W_\Psi f_{b,a} - W_\Psi f_{b,a}^{med}| \geq T_a \\ 0, & \text{if} \quad |W_\Psi f_{b,a} - W_\Psi f_{b,a}^{med}| < T_a \end{cases}, \tag{4}$$

where the threshold $T_a = U * St_a$ determines the presence of an anomaly on the scale a near the point ξ containing in the carrier $\Psi_{b,a}$, U is the threshold coefficient (threshold coefficient $U = 2$ was used for the years of low solar activity and $U = 2.5$ was used for the years with high solar activity), $St_a = \sqrt{\frac{1}{\Phi - 1}\sum_{u=1}^{\Phi}(W_\Psi f_{b,a} - \overline{W_\Psi f_{b,a}})^2}$, $\overline{W_\Psi f_{b,a}}$ and $W_\Psi f_{b,a}^{med}$ are the average value and the median determined in a moving time window of the length Φ.

Anomaly intensity at a time point $t = b$ were estimated as

$$I_b = \sum_a \frac{|P_{T_a}(W_\Psi f_{b,a})|}{\|W_\Psi f_{b,a}\|_2}, \tag{5}$$

where the norm $\|W_\Psi f_{b,a}\|_2 = \sqrt{\sum_{N_a}(P_{T_a}(W_\Psi f_{b,a}))^2}$, N_a is the series length on the scale a.

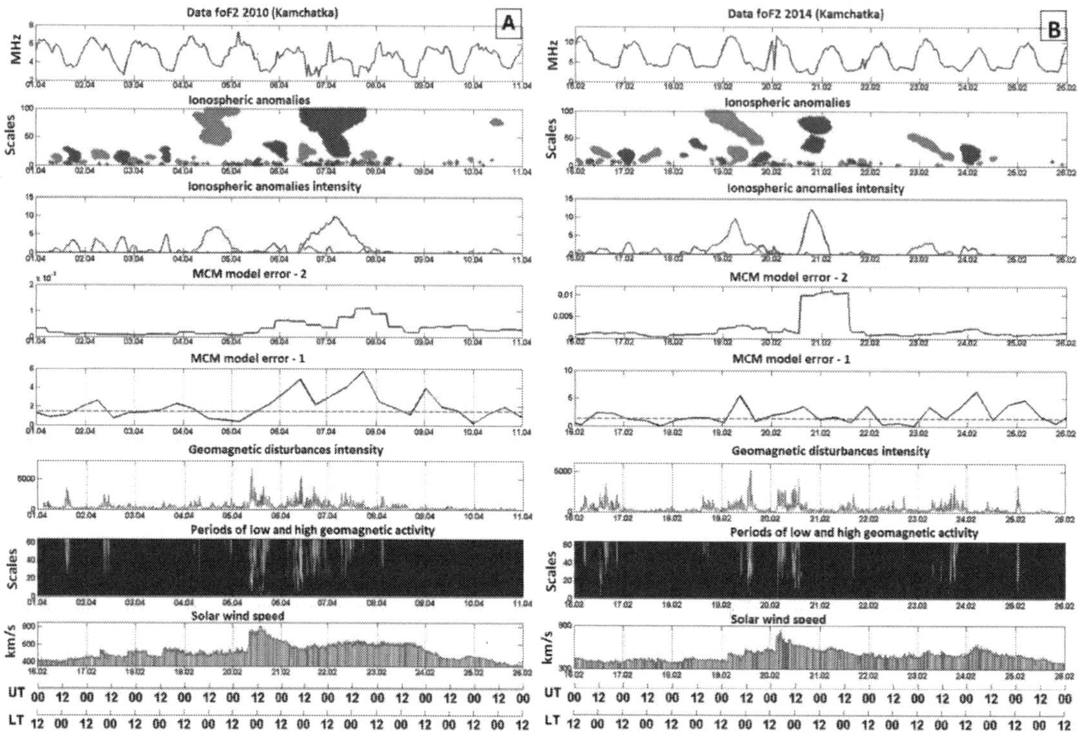

Figure 1. Results of data processing: A - April 01-10, 2010; B - February 16-26, 2014

3 Results of data analysis

For the processing, we applied hourly data of ionospheric parameters (www.ikir.ru) and magnetometer data (http://www.ikir.ru:8180/observ). Fig. 1 shows the results of application of the methods described above on the example of events on April 5-6, 2010 and February 19-20, 2014. A strong magnetic storm occurred within the period under analysis from April 01 to April 10, 2010 (Fig. 1A). The storm was registered on the Earth's surface at 08.26 UT as a sudden commencement (SC). According to the results of processing of ionospheric data, anomalous increases (positive phase of the ionospheric storm, Fig. 1A, marked by red color) were observed in ionospheric parameters on April 4 from 05:00 UT and to the beginning of the magnetic storm. The analysis shows that ionospheric disturbances were the most intensive in the morning (LT). During the storm, the electron concentration decreased significantly and negative phase of ionospheric storm was formed (Fig. 1A, marked by blue color) which lasted for about 30 hours. Based on the results of magnetic data processing, the strongest disturbances of geomagnetic field occurred on April 5-6 during the negative phase of the ionospheric storm. The magnetic storm which occurred on February 19-20, was registered at Paratunka (Kamchatka) site on February 19 at 12:30 UT (Fig. 1B). Anomalous increases in ionospheric parameters were observed before the event from 16:00 on February 18 till 19:00 on February 19 in Kamchatka region (Fig. 1B, marked by red color). Maximum values of ionospheric disturbance intensity were recorded on February 19 at 06:00 UT. By the end of the day on February 19, electron concentration decreased and negative phase of the ionospheric storm was formed which reached the highest intensity from 13:00 UT on February 20 till 04:00 UT on February 21 (Fig. 1B, marked by blue color). During anomalous behavior of the ionosphere, the highest disturbances in geomagnetic field were observed on February from 12:30 UT till 15:30 UT and on February 20 from 03:30 UT till 13:30 UT.

4 Conclusions

The analysis of ionosphere state showed that during increased solar activity, anomalous increases of electron density (more than 2 root mean square deviations relatively the background) may occur before storms in the regions under analysis. The detected increases which lasted from several hours to one day and a half were observed simultaneously at different stations in the North-East of Russia that characterized their large space-time scales. At present, special attention is given to the detection of ionospheric anomalies [19, 20]. The analysis carried out in this work confirms the assumption made in the review [20] that such ionospheric effects are associated with some channel of energy penetration from the interplanetary space and magnetosphere and have solar origin. In this case, the effects may indicate an oncoming geomagnetic storm that is important for space weather forecast [19, 20]. The methods of analysis suggested in the paper showed their efficiency in detection of such effects. The paper was supported by the Russian Scientific Foundation (grant 14-11-00194). We would like to thank the Institutes supporting the stations of cosmic ray registration and magnetic observatories which data were used in the investigations.

References

[1] E. L. Afraimovich et al., Earth, Planets and Space, **52:11**, 1061–1066 (2000)
[2] E. Nakamura, T. Maruyama, Y. Shidama, Journal of the National Institute of Information and Communications Technology, **56**, 391–406 (2000)
[3] K. Watthanasangmechai et al., Earth, Planets and Space, **64**, 473–483 (2012)
[4] O. V. Mandrikova, N. V. Glushkova, I. V. Zhivet?ev, Geomagnetism and Aeronomy, **54:5**, 638–645 (2014)
[5] D. Bilitza, B. W. Reinisc, Advances in Space Research, **42**, 599–609 (2008)
[6] M. V. Klimenko, Earth, Planets and Space, **64:6**, 433-440 (2012)
[7] O. S. Oyekola, P. R. Fagundes, Earth, Planets and Space, **64:6**, 553-566 (2012)
[8] X. Zhao et al., Advances in Space Research, **53:3**, 387–394 (2012)
[9] Y. Kakinami et al., International Journal of Remote Sensing, **31:13**, 3571–3578 (2010)
[10] O. V. Mandrikova et al.,A Journal of Software Engineering and Applications, **5:12B**, 181–187 (2012)
[11] H. Kato et al., Journal of the National Institute of Information and Communications Technology, **56**, 465–474 (2009)
[12] O. V. Mandrikova et al., Earth, Planets and Space, **67**, (2015)
[13] O. V. Mandrikova, N. V. Glushkova, Yu. A. Polozov, Pattern Recognition and Image Analysis, **25:3**, 470–480 (2015)
[14] O. V. Mandrikova et al., Annals of geophysics, **58:5**, (2015)
[15] G. E. P. Box, G. M. Jenkins *Time Series Analysis: Forecasting and Control* (Holden-Day, San Francisco, 1970), 537

[16] O. V. Mandrikova, I. S. Solovev, T. L. Zalyaev, Earth, Planets and Space, **66**, (2014)
[17] O. V. Mandrikova, Yu. A. Polozov, Tsifrovaya obrabotka signalov, **2**, 29–35 (2012), (in Russian)
[18] S. Mallat, *A wavelet tour of signal processing* (Academic Press, London, 1999), 637
[19] D. Buresova, J. Laštovička, Advances in Space Research, **39**, 1298–1303 (2007)
[20] A. D. Danilov, Advances in Space Research, **52:3**, 343–366 (2013)

A neural network system for estimation of the geomagnetic field disturbance

Oksana Mandrikova[1,*] and *Ekaterina* Zhizhikina[1,2,**]

[1] *Institute of Cosmophysical Research and Radio Wave Propagation FEB RAS, 684000, Kamchatskiy kray, Paratunka, Mirnaya st., 7, Russia;*
[2] *Kamchatka State Technical University, 683003, Petropavlovsk-Kamchatsky, Klyuchevskaya st, 35*

Abstract.
The paper presents our software system for estimation the degree of disturbance of the geomagnetic field. The system automatically classifies registered geomagnetic data and determines the state of the geomagnetic field for the current day. The results of approbation of the system showed the prospect of its application in problems of estimation and prediction of space weather. The system allows us to allocate weak disturbances of the geomagnetic field, which may occur before strong magnetic storms.

1 Introduction

The work is devoted to the development of automatic software tools for analysis of geomagnetic data, allocation and estimation of geomagnetic disturbances during periods of increased solar activity and magnetic storms. This paper presents a method developed by the authors and expert software system based on it, which performs classification of recorded variations of the geomagnetic field (H-components of the geomagnetic field are used) and determines its state in an automatic mode.

The recorded variations of the geomagnetic field have a complex structure, therefore classic data analysis methods are ineffective for the task and lead to loss of information [1, 2]. The shortcoming of the used classical methods and approaches is also an insufficient degree of automation, that is very important in the problems of operational data processing and space weather forecast [3, 4]. Papers [2, 5, 6] show that adaptive wavelet decomposition is effective way for analysis of such data. Wavelet transform is used in processing and analyzing complex data structures and it allows us to investigate nonstationary dynamic dependences [1]. In this work we use wavelet transform in conjunction with neural networks. Neural networks are widely used in data classification and pattern recognition[7, 8]. Currently, they are being developed in the field of physics, and, in particular, geophysics [9–11].

2 Description of the method

A characteristic of the geomagnetic field is K-index [3, 12]. We considered three possible states of the geomagnetic field:

*e-mail: oksanam1@mail.ru
**e-mail: ekaterinazh1@mail.ru

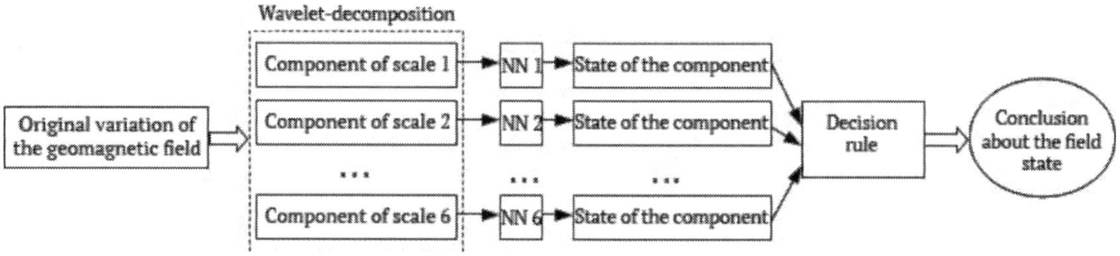

Figure 1. Architecture of the software system

1) "quiet" state (class 1, if the total sum of the daily K-index $\sum K \leq 10$);
2) "weakly disturbed" state (class 2, if $10 < \sum K \leq 18$);
3) "disturbed" state (class 3, if $\sum K > 18$).

Estimation of the geomagnetic disturbance is carried out in accordance with the algorithm described below.

Step 1. Decomposition of the analyzed variation of the geomagnetic field on a different scale components. Using multiresolution wavelet decomposition [13], we obtain the coefficients of the detailing components for the scales $j = -1, -2, ..., -6$ and calculate their absolute values. Obtained coefficients characterize disturbance of the geomagnetic field and in periods of high geomagnetic activity, their absolute values increase significantly [14].

Step 2. Estimation of the state of obtained multiscale components based on radial neural networks. The vectors of the absolute values of coefficients of the scales $j = -1, -2, ..., -6$ are the input vectors of radial neural networks [15]. In the radial layer of each neural network, assesment of proximity measure of the input vector and and example formed during the construction of the network is carried out. The output layers of neural networks determine the probability of belonging of input images to the appropriate class. The process of construction and training of neural networks is described in detail in the paper [14].

Step 3. Estimation of the geomagnetic field state based on team of neural networks. Based on obtained decisions of neural networks, a decision about the state of the geomagnetic field is formed using the following rule:

- if all the components have a "quiet" state, or only one of the components has a "weakly disturbed" state, the geomagnetic field has a "quiet" state (class 1);
- if at least one of the components has a "disturbed" state, the geomagnetic field has a "disturbed" state (class 3);
- in other cases, the field has "weakly disturbed" state (class 2).

Architecture of the software system is shown in Fig. 1.

3 The results of work of the software system

During the study we analyzed geomagnetic field data obtained at the "Paratunka" station (Kamchatka Krai, data registration is carried out by the Institute of Cosmophysical Research and Radio Wave Propagation FEB RAS) for 2013 year, which included 182 quiet variations, 97 weakly disturbed and 40 disturbed. Events of 2013 year with total sum of daily K-index 19 and greater were studied. The system has detected weak disturbances of the field on the eve of the 11 out of 13 such events, while in four cases total per day K-index does not exceed the value of 10. Thus, the system is sensitive to small variations of the geomagnetic field and allows us to detect weak disturbances.

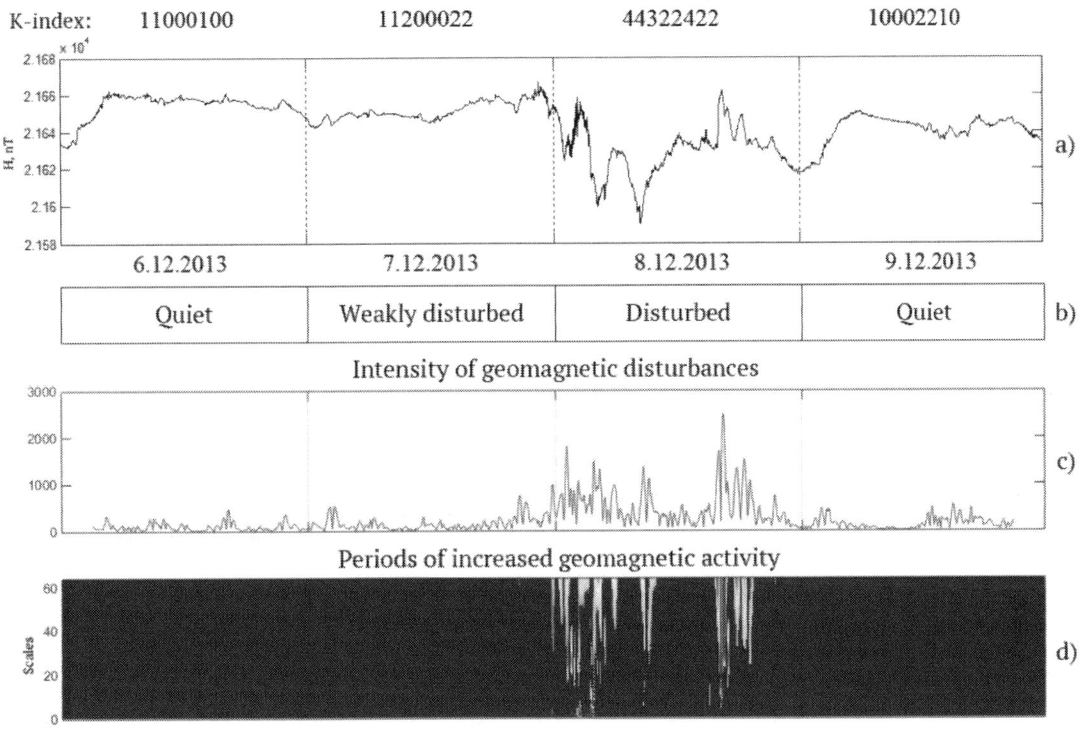

Figure 2. Results of data processing for the period 6-9 December 2013

Fig. 2-4 show the results of data processing. The upper part of Fig. 2-4 shows the values of K-indexes, below the analyzed variations of the geomagnetic field are shown (Fig. 2-4 a) and the results of work of the software system are shown (Fig. 2-4 b). Also, Fig. 2-4 c, d show the results of estimation of the geomagnetic field state obtained on the basis of the algorithm described in [9] and implemented in the software system «Aurora» (this system was developed in IKIR FEB RAS, http://www.ikir.ru:8280/lsaserver/). It can be seen that before the analyzed events, the system detected weak disturbances of the field, although the K-index on the eve of the first 2 events corresponds to the quiet state. The results of applying the algorithm [9] also confirmed the emergence of weak disturbances in the geomagnetic field.

According to the resource [16], a gradual beginning of the storm was recorded at high latitudes on December 7 at 20-00 UT, on December 13 the field was from quiet to very quiet, on June 27 a gradual onset of the storm was recorded at 22-00 UT.

4 Conclusion

The results of the study showed the prospects of application of the developed software system for the analysis of geomagnetic data and estimation of the geomagnetic field state. The system is sensitive to small variations of the geomagnetic field and allows us to automatically allocate weak disturbances that may occur on the eve of magnetic storms. Open access to the system is organized on the website of IKIR FEB RAS (http://www.ikir.ru:8280/lsaserver/).

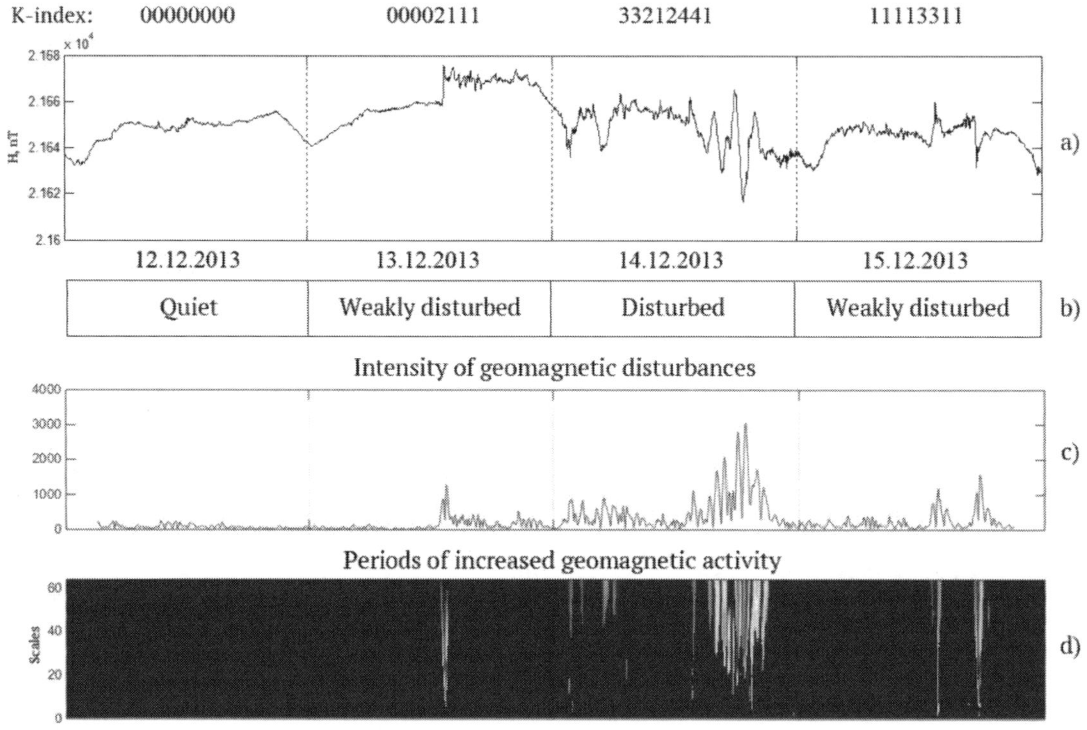

Figure 3. Results of data processing for the period 12-15 December 2013

Development of the method for estimation of geomagnetic field state was performed at the expense of grant funds of the Russian Scientific Fund, project number 14-11-00194. Primary data analysis was performed at the expense of grant funds of RFBR No. 16-55-45007. The authors thank the institutions that support magnetic observatories whose data were used in the study.

References

[1] Zaourar N., Hamoudi M., Mandea M., Balasis G., Holschneider M., Earth Planets Space *65*, 1525-1540 (2013)

[2] Mandrikova O.V., Solovyev I.S., Geppener V.V., Klionskiy D.M., Al-Kasabeh R.T., Digital Signal Processing *23*, 329-339 (2013)

[3] Bud'ko N.I., Zaitsev A.N., Karpachev A.T., Kozlov A.N., Filippov B.P. Edited by Zaitsev A.N. *Kosmicheskaya sreda vokrug nas (In Russian)* (TROVANT, Troitsk, 2006) 232

[4] *Solnechno-zemnye svyazi i kosmicheskaya pogoda, Edited by Petrukovich A.A., chapter 8 in book Plazmennaya geliogeopfizika (In Russian)* (Nauka, Moscow, 2008) 232

[5] Kato H., Takiguchi Y., Fukayama D., Shimizu Y., Maruyama T., Ishii M. Journal of the National Institute of Information and Communications Technology *56*, 465-474 (2009)

[6] Hamoudi M., Zaourar N., Mebarki R., Briqueu L., Parrot M. Geophysical Research Abstracts. EGU General Assembly *11*, EGU2009-8523 (2009)

[7] Savchenko A.V. Computer Optics *37(2)*, 254-262 (2013)

DOI: 10.1051/e3sconf/20161100013

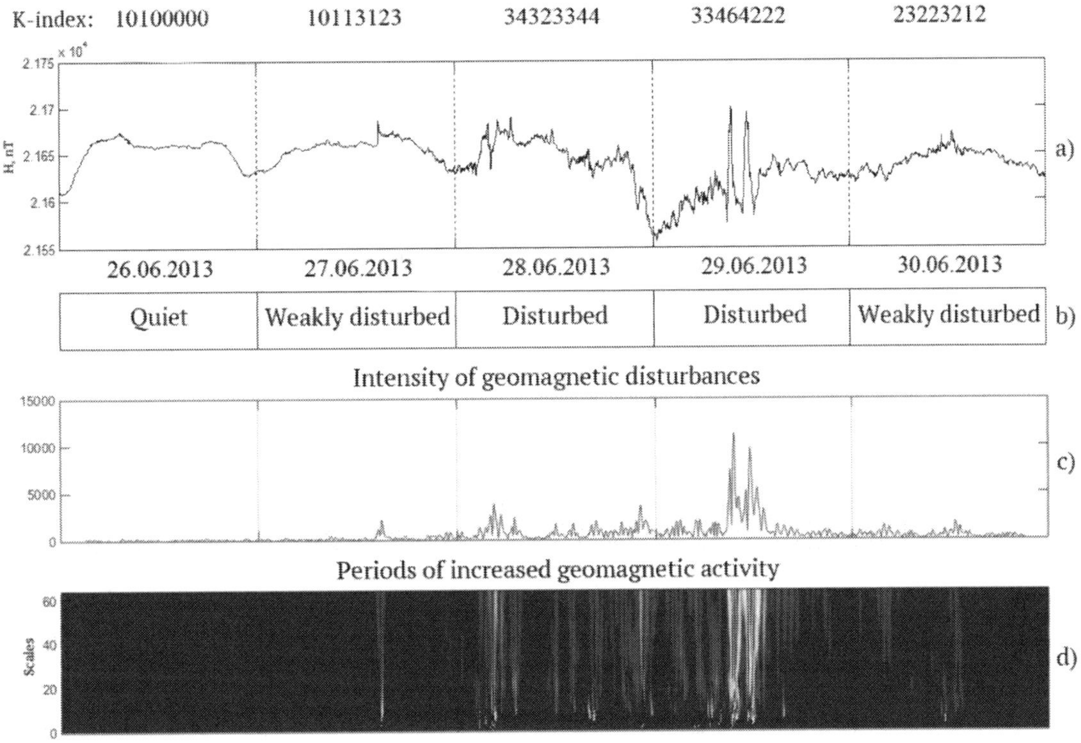

Figure 4. Results of data processing for the period 26-30 June 2013

[8] Soldatova O.P, Garshin A.A. Computer Optics *34(2)*, 252-259 (2010)
[9] Mandrikova O.V., Solovev I.S, Zalyaev T.L. Earth Planet Space *66(1)*, doi:10.1186/s40623-014-0148-0 (2014)
[10] Barkhatova O.M. Solar-Terrestrial Physics *23*, 100-108 (2013)
[11] Uwamahoro J., McKinnell L.A., Habarulema J.B. Annales Geophysicae *30(23)*, 963-972 (2012)
[12] Zabolotnaya N. A Indeksy geomagnitnoy aktivnosti. Spravochnoe posobie (In Russian). (Izd-vo LKI, Moscow, 2007) 88
[13] Chui C. An introduction to wavelets (Academic Press, San Diego,1992) 266
[14] Mandrikova O. V., Zhizhikina E. A. Computer Optics *39(3)*, 420-428 (2015)
[15] Haykin S. Neural Networks: A Comprehensive Foundation. (Prentice Hall, New Jersey, 1999) 823
[16] Obzor kosmicheskoy pogody (URL http://ipg.geospace.ru/space-weather-review/ (date of application 09.05.2016) (In Russian))

Changes of geoacoustic emission directivity at «Mikizha» site associated with earthquakes in Kamchatka

Yuriy Marapulets[1,*], *Alexandra* Solodchuk[1,**], and *Albert* Shcherbina[1,***]

[1] *Institute of Cosmophysical Research and Radio Wave Propagation FEB RAS, Russia, 684034, Kamchatka region, Elizovskiy district, Paratunka, Mirnaya str., 7*

Abstract. To investigate high frequency geoacoustic emission, a receiving system based on a compound vector receiver was installed in Kamchatka. It allows the authors to determine the direction of sound wave arrival. In the result of data analysis for the period from August 2008 to January 2016, it was determined that anomalies of geoacoustic emission directivity occur during the majority of the earthquakes with $K_s > 9.0$ in the South of Kamchatka.

1 Introduction

Investigations of geoacoustic emission in the frequency range from 0.1 Hz to 10 kHz are carried out at the Institute of Cosmophysical Research and Radio Wave Propagation, Kamchatka peninsula. Analysis of geoacoustic signal shows that it consists of a series of relaxation pulses of different amplitude and duration with shock excitation and the carrier frequency from units to tens and more kilohertz. The pulse repetition frequency (acoustic activity $\Omega(t)$) is determined by rock deformations and may vary within a wide range, from single signals on a time interval of several of seconds during calm periods to tens and even hundreds per a second during intensification of plastic process, in particular before earthquakes. This acoustic effect manifests itself the most vividly in the kilohertz frequency range [1], [2]. Some cases have been registered when acoustic activity increased suddenly some days before an earthquake. Clear maxima were observed on the diagram of its azimuthal distribution [3], [4]. Thus, it is reasonable to analyze the changes of high-frequency geoacoustic emission directivity during earthquake preparation in Kamchatka.

2 Methods of investigations

Investigations of the directivity properties of high-frequency geoacoustic emission are carried out by a hardware-software complex installed at "Mikizha" observation site of IKIR FEB RAS. Registration of geoacoustic data and estimations of a sound wave arrival direction are realized by a receiving hydroacoustic system constructed on the base of a combined receiver manufactured by ZAO "Geoakustika" and mounted by the bottom of a natural lake [5].

*e-mail: marpl@ikir.ru
**e-mail: aleksandra@ikir.ru
***e-mail: albert@ikir.ru

To analyze the changes of geoacoustic emission directivity during earthquakes, we chose 4 groups of earthquakes from the seismic catalogue of Kamchatka Branch of Geophysical Service RAS. The earthquakes occurred during the period from August 2008 to January 2016 with the energy classes of $9.0< K_s \leq 10$, $10< K_s \leq 11$, $11< K_s \leq 12$ and $K_s >12$ and at the epicentral distances of $R \leq 120$ km, $R \leq 190$ km, $R \leq 240$ km and $R \leq 350$ km, respectively. Such distances were chosen taking into account the results of investigations [6] and corrected based on [7]. The earthquakes which occurred from February 10 to July 09, 2013 were excluded from the analysis due to the absence of geoacoustic data during the mentioned period. Thus, 276 seismic events were chosen.

During calm weather conditions (weakly changing pressure, absence of precipitation and wind of more than 6 m/s) within the interval of ± 5 days close to an earthquake, acoustic activity $\Omega(t)$ and its azimuthal distribution $D(\alpha, t)$ were under consideration. It was believed that anomalies associated with a seismic event are observed in the directivity of geoacoustic emission when in the time interval mentioned above, the pulse repetition frequency and its distribution in one or in several directions exceed the values of the corresponding background levels by 2.5 and more times. Background levels are chosen according to the results of preliminary analysis of long-term observation series of acoustic activity.

3 Results of investigation and discussion

In the result of the analysis it was ascertained that there are anomalies in geoacoustic emission directivity which occur close to 219 earthquakes out of 276 (79%). Anomalies which could be associated with several seismic events were not included into further consideration because it was difficult to determine the impact of each event into the investigated effects. Fig. 1 shows an example of such an anomaly.

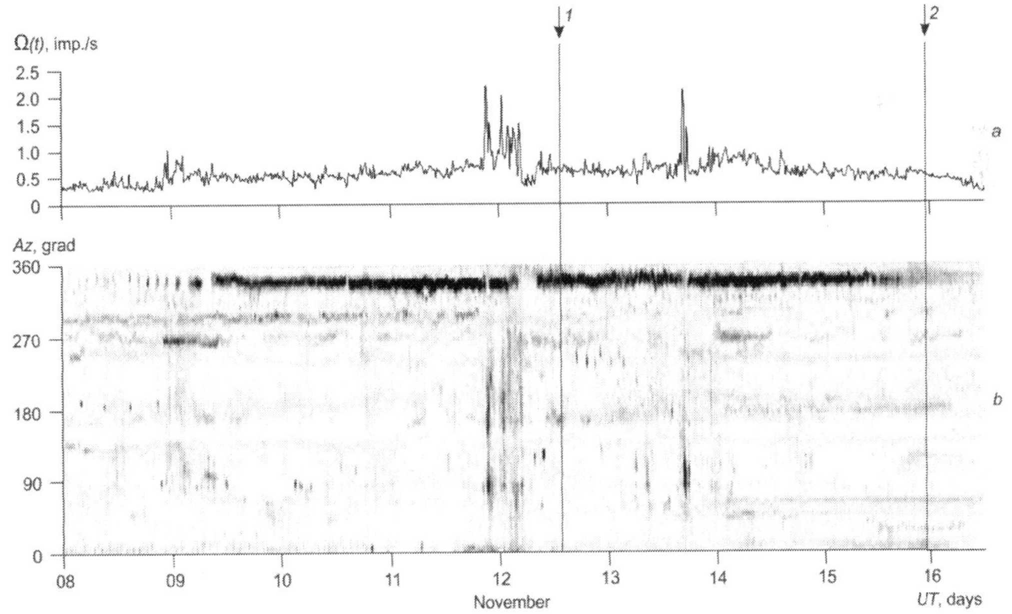

Figure 1. Acoustic activity $\Omega(t)$ (*a*) and its azimuthal distribution $D(\alpha, t)$ (*b*), registered close to the earthquakes on 12.11.2015 at 13:54 UT (51.39°N, 158.90°E) with K_s=12.3 and R=180 km (*1*) and on 15.11.2015 at 23:17 UT (52.39°N, 159.36°E) with K_s=11.7 and R=100 km (*2*). *Az* — azimuth. Arrows continued as lines indicate the times of earthquakes.

Thus, anomalies of directivity observed close only to one earthquake were analyzed. They were registered in 73 cases. It was ascertained that anomaly duration varies within the range from 10 to 60 hours (Fig. 2*a*). The increased activity is mainly registered from the directions corresponding to the amithuths of 50°, 155°, 225°, 330° (Fig. 2*b*). Similar azimuths were earlier detected during the investigations of geoacoustic emission directivity [4].

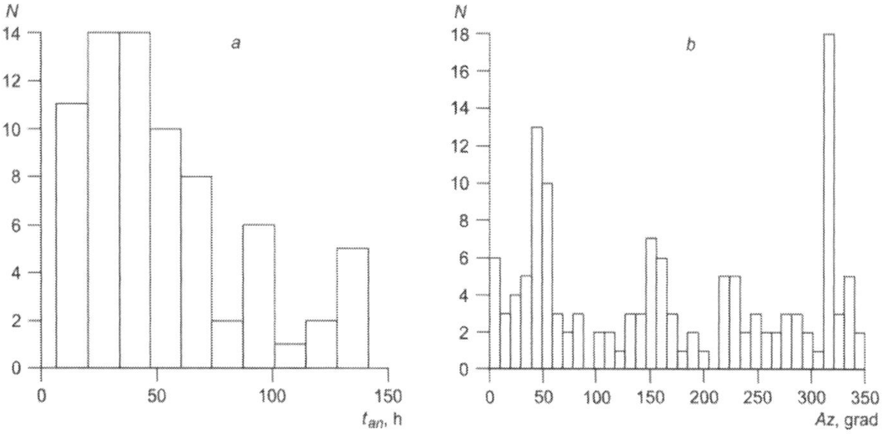

Figure 2. Histograms of duration t_{an} (*a*) and of azimuths Az (*b*) of geoacoustic emission anomalies. N — the number of anomalies.

Analysis of the duration of geoacoustic emission directivity anomalies close to seismic events showed that they occur 10–30 h before earthquakes and last for 6–30 h after them in most cases.

Fig. 3 illustrates the dependences of anomaly duration on earthquake class (*a*) and on epicentral distance (*b*). Correlation coefficients of these values are -0.04 and 0.01 at the significance levels of 0.7 and 0.9, respectively. Thus, we have not discovered any relation between the parameters under investigation.

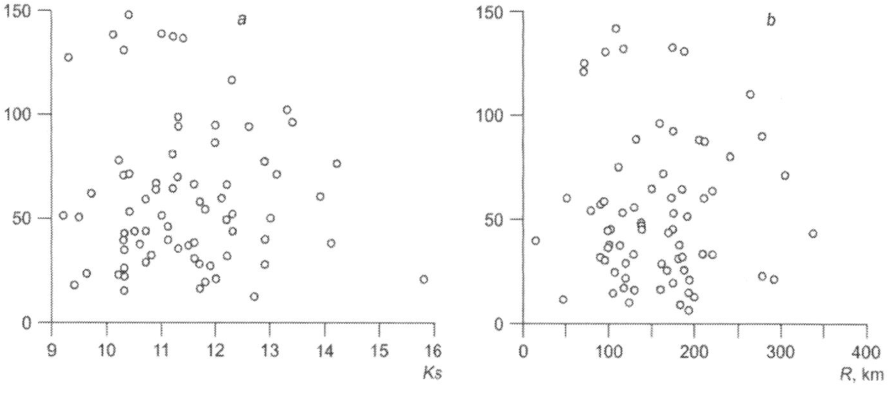

Figure 3. Dependences of anomaly duration t_{an} on earthquake class K_s (*a*) and epicentral distance R (*b*).

As an example, Fib. 4 and Fig. 5 show anomalies of geoacoustic emission directivity associated with earthquakes.

Fig. 4 illustrates an anomaly registered 30 h before the earthquake with the energy class of K_s=10.3, which occurred 17.12.2012 at 03:10 UT (51.87°N, 159.12°E) at the epicentral distance of R=140 km (*1* on the graph), and ended in 17 h after it (anomaly duration was 47 h). On the graph of integral activity (Fig. 4*a*) sharp increase of geoacoustic pulse repetition frequency is observed (with background level increase by more than 4 times). Activations of the directions in the range of 40°–50° correspond to it (Fig. 4*b*).

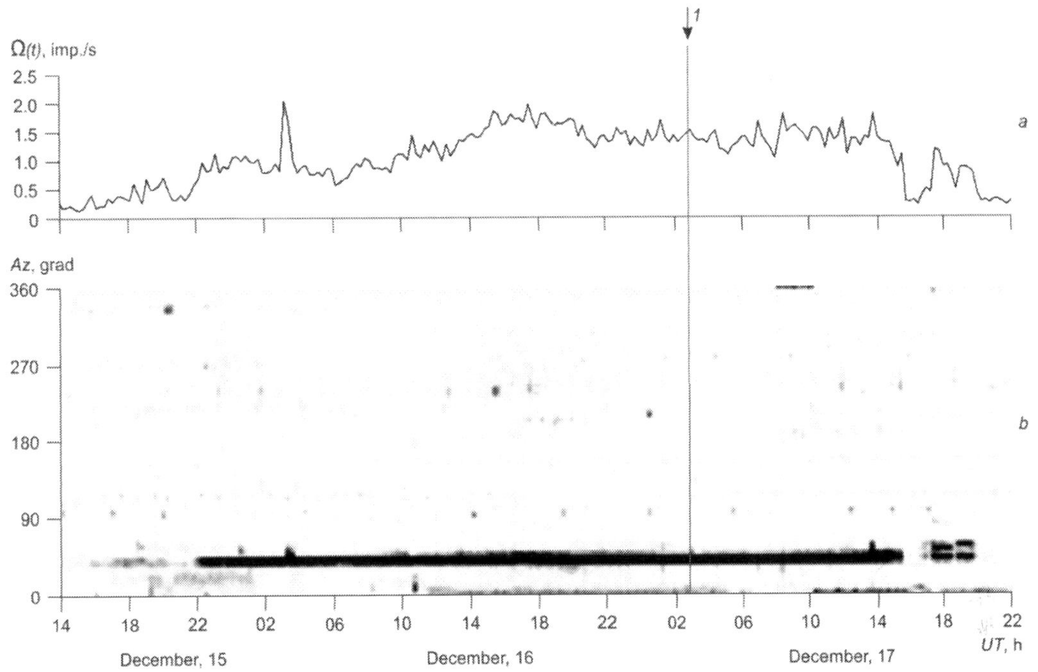

Figure 4. Acoustic activity $\Omega(t)$ (*a*) and its azimuthal distribution $D(\alpha, t)$ (*b*) close to the earthquake with the energy class of K_s=10.3 (*1*). *Az* — azimuth. Arrows continued as lines indicate the times of earthquakes.

Fig. 5 shows anomalies of geoacoustic emission directivity associated with the earthquakes with the energy class of K_s=15.7 which occurred on 30.01.2016 at 03:25 UT (53.85°N, 159.03°E) at the epicentral distance of R=110 km (*1* on the graph). Its strongest aftershocks were registered on 30.01.2016 at 03:42 UT (53.75°N, 159.07°E) with K_s=11.6 and R=100 km (*2* on the graph) and at 06:53 UT (53.87°N, 159.21°E) with K_s=10.5 and R=120 km (*3* on the graph). The anomaly with the direction range of 330°–350° (Fig. 5*b*), during which the exceedance of the acoustic activity background level by more than 5 times was observed (Fig. 5*a*), began 15 h before the earthquake with K_s=15.7 and ended about 30 minutes before it. Almost simultaneously with it, a less increase of activity was registered in azimuth 280° (Fig. 5*b*). It continued after the earthquake and was likely associated with aftershocks.

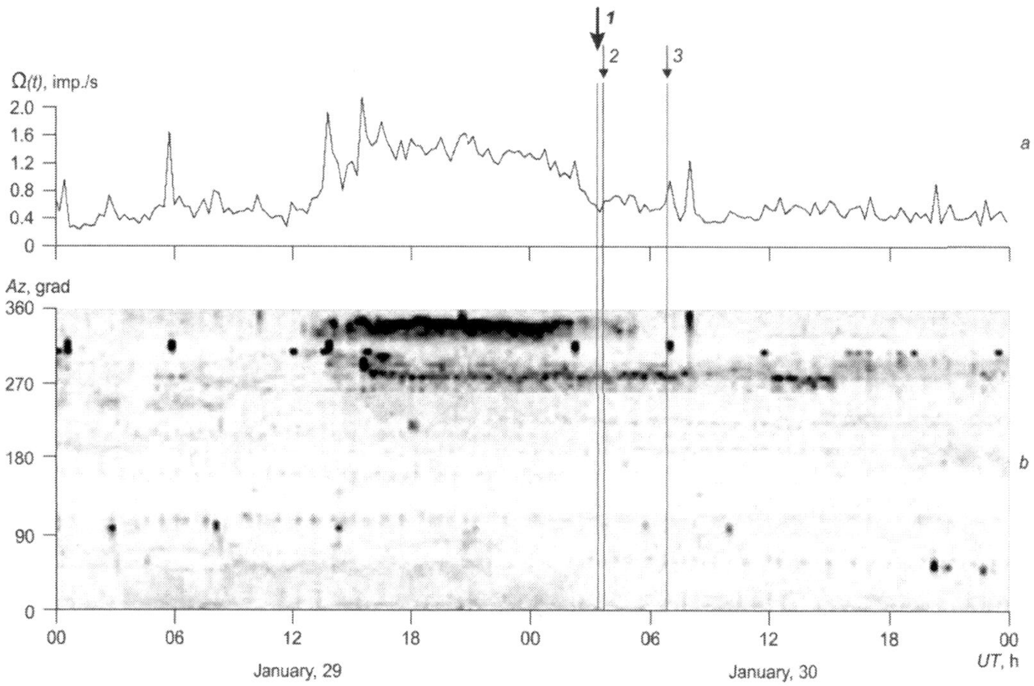

Figure 5. Acoustic activity $\Omega(t)$ (*a*) and its azimuthal distribution $D(\alpha, t)$ (*b*) close to the earthquake with K_s=15.7 (*1*) and its aftershocks (*2* and *3*). *Az* — azimuth. Arrows continued as lines indicate the times of earthquakes.

Conclusions

Thus, in the result of the analysis of the high-frequency geoacoustic emission directivity change, we determined anomalies associated with earthquakes. They appear as sharp increases of geoacoustic pulse repetition frequency in some directions. Such effects occur 10–30 h before earthquakes and last for 6–30 h after it in most cases.

References

[1] Dolgikh G. I., et al. Doklady Earth Sciences **413:2**, 281–285 (2007)

[2] Marapulets Yu. V., Shevtsov B. M., *Mezomasshtabnaya akusticheskaya emissiya [Mesoscale acoustic emission]* (Dal'nauka, Vladivostok, 2012) 126 p.

[3] Shevtsov B. M., Marapulets Yu. V., Shcherbina A. O. Doklady Earth Sciences **430:1**, 67–70 (2010)

[4] Marapulets Yu. V., Shcherbina A. O. Solnechno-zemnye svyazi i fizika predvestnikov zemletryaseniy, VI mezhdunarodnaya konferentsiya, ed. Shevtsov B. M., sb. dokl. , 316–319 (2013)

[5] Marapulets Yu. V., Shcherbina A. O. Pribory **9(183)**, 14–19 (2015)

[6] Kuptsov A. V. Izvestiya. Physics of the Solid Earth **41:10**, 825–831, (2005)

[7] Dobrovol'skiy I. P. *Mekhanika podgotovki tektonicheskogo zemletryaseniya [Mechanics of tectonic earthquake preparation]* (IFZ AN SSSR, M, 1984) 189 p.

Statistics of occurrence of pre-seismic anomalies in geoacoustic emission and in atmospheric field

Mikhail Mishchenko[1,*]

[1] *Institute of Cosmophysical Researches and Radio Wave Propagation FEB RAS, Paratunka, Kamchatsky Kray, Russia*

Abstract. The paper presents the statistics of occurrence of anomalous disturbances in high-frequency geoacoustic emission of near surface sedimentary rocks and in atmosphric electric field by the ground surface before earthquakes. Long-term continuous series of geoacoustic emission measurements in 2003-2012 at "Mikizha" observation site and atmospheric electric field data obtained within summer-autumn period in 2006-2008 at "Mikizha" site and in 2009-2012 at "Karymshina" observation site were used for the analysis. Anomalous disturbances of the emission and of the field were compared with the earthquake catalogue of Kamchatka Branch of Geophysical Service RAS.

1 Introduction

Measurements of geoacoustic emission of the near surface sedimentary rocks (GAE) and of atmospheric electric field by the ground surface (AEF) are carried out in Kamchatka in the conditions of active seismic processes. Joint geoacoustic and atmospheric-electric measurements [1] indicate the fact that anomalous disturbances of these geophysical fields occur during seismically calm periods and at the final stage of earthquake preparation and are associated with the dynamics of the deformation process of the near surface sedimentary rocks. Thus, statistics of occurrence of such disturbances before earthquakes both in geoacoustic emission and in atmospheric electric field is of interest.

2 Methods of investigation

To analyze the preseismic disturbances, we used a long-term continuous series of geoactoustic measurements of the period from 2003 to 2013 from "Mikizha" site and AEF measurement data carried out during the summer-autumn period in 2005-2008 at "Mikizha" site and in 2009-2012 at "Karymshina" site. Anomalous disturbances of geoacoustic emission and those of atmospheric electric field were compared with the regional earthquake catalogue of Kamchatka Branch of Geophysical Service RAN (http://www.emsd.ru/ts/). Only GAE and AEF disturbances registered in fair weather conditions (slightly changing pressure, absence of rain and wind of more than 6 m/s) were under consideration.

* e-mail: micle@ikir.ru

3 Results of investigation and discussion

Analysis of the data of geoacoustic observations in seismically active regions showed that anomalous disturbances of geoacoustic emission are sometimes registered before strong earthquakes. They appear as sudden increases of acoustic pressure in kilohertz frequency range Ps [2, 3]. An example of such a disturbance of geoacoustic emission is shown in Fig. 1a. 266 disturbances of GAE were detected for the period of 2003–2007. From their analysis, the following criteria for disturbance detection were selected: Ps amplitude averaged within 5-minute interval should exceed the GAE background value by more than 4 times with the duration of not less than 15 min. Averages of Ps amplitude for the calmest day period within the closest ten days before a disturbance were taken as GAE background level [4].

Analysis of the data of atmospheric-electric observations in seismically active regions showed that anomalous disturbances of electric field are sometimes registered before strong earthquakes. As such, we usually consider bay decreases of field potential gradient V' up the change of a sign followed after a while by its recovery almost to the previous level [5–7]. An example of such a disturbance of AEF is shown in Fig. 1b.

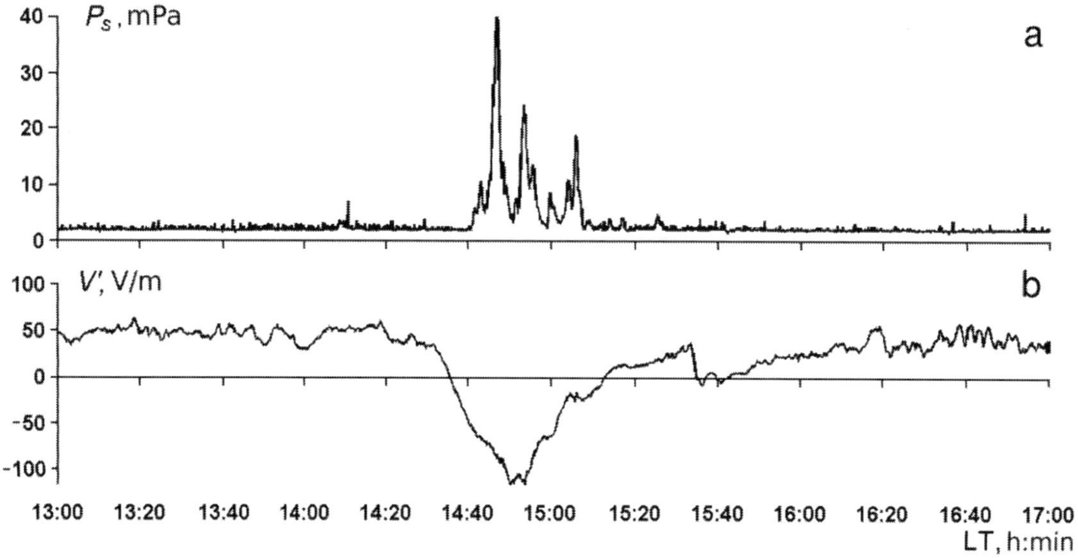

Figure 1. Example of simultaneous disturbances of acoustic pressure Ps in kilohertz frequency range (a) and that of potential gradient V' (b) in the experiment on August 22, 2006.

Taking into account that atmospheric-electric measurements were carried out during the summer-autumn periods and lasted for only about three month a year, we did not receive enough data on preseismic disturbances of AEF to determine the criteria for their detection. Thus, the criteria were chosen based on the analysis of preseismic characteristics of AEF in Kamchatka. The analysis was carried out in the paper [7] for a continuous period of 1997–2002. The minimum duration of AEF disturbances is 15–20 min, and the minimum amplitude of bay decreases of the potential gradient V' is −100 V/m.

Geoacoustic and atmospheric-electric disturbances occur almost simultaneously and are comparable in duration. Thus, in the further analysis, just like for GAE disturbances, we distinguished preseismic AEF disturbances with the duration of more than 15 min and the minimum decrease of the potential gradient V' equal to -100 V/m that agrees with the results obtained in the paper [7].

Taking into account the criteria for GAE disturbance detection, we analyzed geoacoustic data series from "Mikizha" site registered within 2003–2012. Anomalous disturbances of geoacoustic emission registered in fair weather conditions were detected. From the regional seismic catalogue, we chose two groups of earthquakes with the energy classes $11.0 < Ks \leq 12.0$, $Ks > 12.0$ and epicentral distances to the site $D \leq 240$, $D \leq 350$, respectively. Such distances were selected according to the analysis of GAE disturbance data [3] and corrected using the model studies [8] as the base. In the result of modeling [4], the time interval from two days and a half to 5 hours before an earthquake was chosen to search for GAE disturbances. The chosen earthquakes were compared with GAE anomalous disturbances. Presence of GAE disturbances before 107 earthquakes was considered. The results are shown in Table 1.

Table 1. Results of comparison of earthquakes with GAE anomalous disturbances

Earthquakes						
Energy class	$11.0 < Ks \leq 12.0$			$Ks > 12.0$		
Year	Total	with GAE disturbances	%	All	with GAE disturbances	%
2003	7	5	71.4	3	3	100
2004	4	2	50	6	6	100
2005	1	0	0	2	2	100
2006	7	5	71.4	4	3	75
2007	6	5	83.3	3	3	100
2008	5	4	80	6	3	50
2009	8	6	75	1	0	0
2010	5	1	20	7	3	42.9
2011	17	9	52.9	3	3	100
2012	9	5	55.6	3	2	66.7
Total	69	42	60.9	38	28	73.7

According to Table 1, GAE disturbances were observed before 42 out of 69 earthquakes with the characteristics $11.0 < Ks \leq 12.0$ and $D \leq 240$ km and before 28 out of 38 earthquakes with $Ks > 12.0$ and $D \leq 350$ km.

The results of comparison of geoacoustic emission with the earthquakes with energy class $Ks >$ 9.0 showed that 36.8% of GAE disturbances occured before earthquakes. Thus, the percentage of false alarms is 63.2%.

In terms of the criteria formulated for detection of AEF disturbances, we analyzed the series of atmospheric-electric data for the periods of summer-autumn measurements at "Mikizha" site (2006, 2007 and 2008) and at "Karymshina" site (2009, 2011 and 2012). Anomalous disturbances of atmospheric electric field by the ground surface were detected during fair weather conditions described above.

Just like for GAE, two groups of earthquakes with energy classes $11.0 < Ks \leq 12.0$ and $Ks >$ 12.0 and epicentral distances $D \leq 240$ and $D \leq 350$ km were chosen. As long as geoacoustic emission is the indicator of change of the near surface rock deformation state, time intervals of two days and a half before earthquakes were chosen according to [4] to search for AEF disturbances. The selected earthquakes were compared with AEF anomalous disturbances. For "Mikizha" site, presence of AEF disturbances was considered before 19 earthquakes and for "Karymchina" site we analyzed 11 earthquakes. The results of comparison are shown in Table 2.

Table 2. Results of comparison of earthquakes with AEF anomalous disturbances

Earthquakes					
Energy class		$11.0 < Ks \leq 12.0$		$Ks > 12.0$	
Observation site	Year	Total	with AEF disturbances	All	with AEF disturbances
"Mikizha"	2006	4	1	4	1
	2007	3	1	0	0
	2008	3	0	2	1
"Karymchina"	2009	2	2	0	0
	2011	3	1	1	1
	2012	4	1	1	0
	Total	19	6	8	3

According to Table 2, AEF disturbances were observed before 2 earthquakes out of 10 with characteristics $11.0 < Ks \leq 12.0$ and $D \leq 240$ km and before 2 earthquakes out of 6 with $Ks > 12.0$ and $D \leq 350$ km at "Mikizha" site. At "Karymshina" site, AEF disturbances were observed before 4 out of 9 earthquakes with the characteristics $11.0 < Ks \leq 12.0$ and $D \leq 240$ km and before 1 out of 2 earthquakes with $Ks > 12.0$ and $D \leq 350$ km.

4 Conclusions

Based on the analysis of the observations of geoacoustic emission and of atmospehric electric field, criteria for detection of their anomalous disturbances were suggested. They allowed us to distinguish disturbances preceding the earthquakes in high-frequency geoacoustic emission and in atmospheric electric field by the ground surface in Kamchatka. The number of such preseismic disturbances in the electric field is much less than that in GAE since atmospheric-electric measurements were carried out only during summer-autumn periods and lasted for about three month a year.

References

[1] O. P. Rulenko, Yu. V. Marapulets, M. A. Mishchenko, Journal of Volcanology and Seismology, **8:3**, 183–193 (2014)
[2] V. A. Morgunov et al., Volcanology and Seismology, **13:4**, 516–518 (1992)
[3] A. V. Kuptsov, Izvestiya. Physics of the Solid Earth, **41:10**, 825–831 (2005)
[4] M. A. Mishchenko, Bulletin KRASEC. Phys.&Math. Sci., **2:1**, 56–64 (2011)
[5] O. P. Rulenko, A.V. Ivanov, A. V. Shumeyko, Doklady Earth Sciences, **326:6**, 980–982 (1992)
[6] O. P. Rulenko, Volcanology and Seismology, **22:4**, 435–451 (2001)
[7] S. E. Smirnov, Geomagnetism and Aeronomy, **45**, 265–269 (2005)
[8] I. P. Dobrovol'skiy *Mekhanika podgotovki tektonicheskogo zemletryaseniya* (IFZ AN SSSR, Moscow, 1984), 189 (in Russian)

The influence of the droplet clouds microstructure on the polarization characteristics of a double scattering lidar signal

Yevgeniy Nee, *Valentina* Bryukhanova, and *Anton* Doroshkevich

Tomsk State Universities, Radiophysics Department, Russian Federation

Abstract. The results of calculation of the ratio of polarization degrees of the double scattering lidar return from droplet clouds with different microstructure at sensing by circularly and linearly polarized radiation are given in this report. The influence of the droplet size on ellipse parameters of linearly polarized radiation are discussed.

1 Introduction

Determination of the clouds microstructure is significant scientific and practical interest. This is explained by the clouds make significant impact on the radiation balance of the Earth. The lidar allows us to study clouds without changes of them in near real-time with high spatial resolution [1, 2]. Most often, the interpretation of the lidar data is based on the lidar equation, which was got for the assumption that sensing of the atmosphere is limited by the contribution of photons scattered on the receiving system direction only the once. This approach is justified for the aerosol formation with small optical depth such as low-density haze and high-level clouds. However, the multiple scattering makes significant contribution to the lidar return at the study of many types of low-level and medium-level clouds, fogs, dense hazes.

Experimental studies of brightness and polarization characteristics of the multiple backscattering radiation [3] showed that the intensity of the multiple scattering signal is distributed over the full field-of-view of the lidar receiving system. Single scattering lidar return is formed in the space determined by the radiation pattern of the lidar source and doesn't depend on the lidar field-of-view. It is showed in the paper [4], the value of multiple scattering lidar return associates with the generalized parameter $\eta = H\sigma \mathrm{tg}(\theta/2)$, which is determined by scattering coefficient (σ) and cross-section of the scattering volume ($H\mathrm{tg}(\theta/2)$). So, the multiple scattering lidar return is formed in the scattering volume limited by the field-of-view of lidar receiving system: more the receiving system field-of-view, more the cross-section of scattering volume and, therefore, more the multiple scattering contribution in the lidar return. Thus, the changes of power of the lidar return by increasing the receiving system field-of-view of lidar occurs only due to the increase of multiple scattering signal. Therefore, the component of multiple scattering can be recovered from the lidar return by using lidar with multiple field-of-view [5-8].

The use of the stop blocked single scattering radiation in the lidar receiver for registration of the multiple scattering lidar return was offered in the paper [5]. In this case, the photosensitive element receives only the multiple scattering radiation. Diaphragms with different diameter which was set in the receiving system by turns are used for the decomposition of a lidar return into single and multiple scattering components in [6, 7]. Using a CCD-camera [8] set in the focal plane of a receiving lens, is equivalent to carrying out the experiment with several receiving channels with different field-of-view.

The multiple scattering coaxial lidar is developed for a detailed study of the lidar return structure from tropospheric clouds by the OES&RS department of TSU. The CCD camera will be used as a receiver [9].

2 Double scattering lidar return

In many practically significant cases the lidar return can be described based on double scattering approximation with sufficient accuracy. Model of the double scattering lidar return was reported in the most details by Kaul and Samokhvalov [10].

The polarization state of lidar return can be conveniently represented in the form of the Stokes vector. The expression describing the Stokes vector of double scattering lidar return looks as follows [11]:

$$\vec{S}(r)=\vec{S}^{(1)}(r)+\vec{S}^{(2)}(r)=\frac{P^{(1)}(r)}{X(\pi)}\left[\hat{M}(\pi)\vec{S}_0+2\sigma_0 r\left(\vec{I}_1+\vec{I}_2\right)\right], \tag{1}$$

where

$$\vec{I}_1=\sigma_0(r-H)\int_0^{\gamma_1}\int_0^{2\pi}\vec{G}(\varphi,\gamma,\pi-\gamma)\mathrm{tg}\frac{\gamma}{2}d\varphi d\gamma,\quad \vec{I}_2=\int_{\gamma_1}^{\pi/2}\int_0^{2\pi}\vec{G}(\varphi,\pi-\gamma,\gamma)d\varphi d\gamma,$$

$$\gamma_1=2\,\mathrm{arctg}\frac{r\,\mathrm{tg}\frac{\theta_0}{2}}{r-H},$$

$$\vec{G}(\varphi,\gamma,\pi-\gamma)=\hat{R}(\varphi)\hat{M}(\pi-\gamma)\hat{M}(\gamma)\hat{R}(\varphi)\vec{S}^0$$

is the integral matrix vector-parameter defined by the operator of a references plane rotating on the angle φ concerning base

$$\hat{R}(\varphi)=\begin{pmatrix}1 & 0 & 0 & 0\\ 0 & \cos 2\varphi & \sin 2\varphi & 0\\ 0 & -\sin 2\varphi & \cos 2\varphi & 0\\ 0 & 0 & 0 & 1\end{pmatrix},$$

$\hat{M}$ – scattering phase matrix, which contains the maximum possible information on the scattering medium [12]. In the general case it comprises 16 components

$$\hat{M}=\begin{pmatrix}f_{11} & f_{12} & f_{13} & f_{14}\\ f_{21} & f_{22} & f_{23} & f_{24}\\ f_{31} & f_{32} & f_{33} & f_{34}\\ f_{41} & f_{42} & f_{43} & f_{44}\end{pmatrix}.$$

Elements of this matrix describe by the combination of equations for the amplitude of the scattered radiation field. Scattering phases matrix depends on the complex refractive index, particle size and shape [13].

3 The dependence of the polarization degree of the double scattering lidar return from the water content of the cloud

We use of droplet cloud models [14] been typical of the Kamchatka region [15] for calculation of polarization characteristics of the double scattering lidar return (tab.1). Optical characteristics of these clouds were calculated by using the program PolyMie [16].

Table 1. Parameters of the droplet clouds microstructure.

r_m, µm	µ	N, см$^{-3}$	δ, g/m^3
4,48	10,6	350	0,220
4,94	5,8	58	0,070
4,94	7,2	96	0,100
5,28	8,5	80	0,092
5,46	2,8	59	0,194
5,72	6,4	103	0,180
5,95	3,4	39	0,133
5,95	22,2	165	0,188

Here, r_m – modal radius of the particle size distribution, µ – half-width parameter, N – particle number concentration, δ –cloud liquid water content.

Analysis of calculation results shows that the ratio of polarization degrees of double scattering lidar return at sensing circularly and linearly polarized radiation doesn't depend on the lidar receiving system field-of-view (Fig.1). This allows us to speed up the processing of the experimental data by reducing amount of data read out from the CCD-camera and used for the interpretation of experimental data.

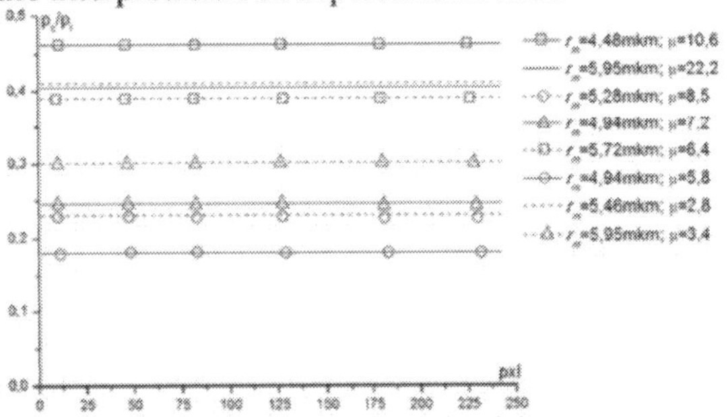

Fig. 1. The dependence of the ratio of polarization degrees of the double scattering lidar return at sensing circularly and linearly polarized radiation on the field-of-view of the lidar receiving system.

A more detailed research of the dependence of the ratio of polarization degrees of the double scattering lidar return on microstructure parameters of a cloud showed that the increase of the polarization degree ratio occurs with the increase of liquid water content of clouds (Fig. 2).

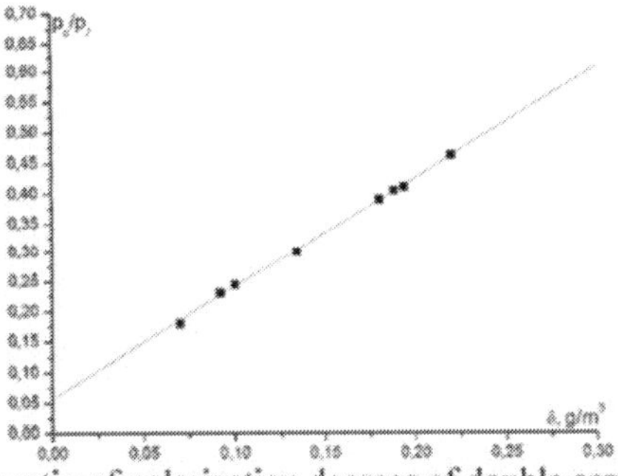

Fig. 2. The dependence of the ratio of polarization degrees of double scattering lidar return at sensing circularly and linearly polarized radiation on the water content of cloud been typical of the Kamchatka region.

In the general case the dependence of the ratio of polarization degree at sensing circularly and linearly polarized radiation on the liquid water content of cloud is nonlinear [17] for all models of clouds reported in [14]. The slope angle of the approximating function for cloud models described above is 1.08±0.02. As you can see cloud models been typical of Kamchatka region is in good agreement with the present trend.

4 The influence of particle size on polarization ellipse parameters of the double scattering lidar return

The study of the influence of sensing droplet cloudy microstructure on the polarization state of the double scattering lidar return calls to analyze the transformation of polarization state of sounding radiation.

Previously [18, 19], we calculated the distribution of the polarization degree and the elements of the normalized Stokes vector over the focal plane of the receiving system. These calculations were based on the expression (1). According to results the pattern of intensity distribution over the detection plane depends on the polarization state of sounding radiation. There are areas of prevailing depolarization and full polarization. If laser pulse is linearly polarized the areas of prevailing depolarization (45°, 135°, 225° и 315°) and total polarization (0°, 90°, 180° и 270°) will be observed. If laser pulse is circular polarized the distribution pattern is equal for all azimuthal angles [19].

In order to analyze the structure features of the polarization degree distribution we studied the distribution of the ellipticity ε and the semi-major axis orientation of the ellipse of polarization φ over the detection plane of the receiving system.

$$\varepsilon = \frac{1}{2}\arcsin\left(\frac{S_4}{\sqrt{S_2^2+S_3^2+S_4^2}}\right), \qquad \varphi = \frac{1}{2}arctg\left(\frac{S_3}{S_2}\right).$$

According to (1) the Stokes vector of the double scattering lidar signal is determined by the size of the scattering volume, which depends on the distance to an aerosol formation of H and the receiving system field of view θ0, and by optical properties of this volume (scattering coefficient σ and phase scattering matrix M).

We used continental droplet cloud models for a scattering media. To reduce the influence of geometrical and optical parameters of aerosol formation on the study results the selection criterion was the similar value of the cloud base H and the scattering coefficient σ. Chosen models lie in the line section in Fig. 2. The droplet number concentration N, mode radius r_m, half-width parameter of the gamma particle size distribution μ and liquid water content δ of studied aerosol formation are shown in table 2.

Table 2. The parameters of droplet cloud models.

№	H, м	N, см$^{-3}$	r_m, мкм	μ	δ, г/м^3	σ, км$^{-1}$
1	840	680	2,68	13,0	0,08	19,59
2	750	396	2,83	3,3	0,15	21,00
3	908	480	3,49	14,8	0,12	20,01
4	1000	250	4,38	6,2	0,20	20,19
5	980	297	4,41	13,3	0,16	19,27
6	950	165	5,95	22,2	0,19	20,69

The particle size distributions of selected cloud models are presented in Fig. 3. Models 1 and 2 and also 4 and 5 have similar values of the modal radius but different half-width of the particle size distribution. On the other hand, the models 2 and 4 and also 5 and 6 have different modal radius but identical half-width of the distribution. Thus, these models allow us to evaluate the influence of the modal radius and the half-width of the particle size distribution on the distribution parameters of polarization characteristics of the double scattering lidar return in the detection plane.

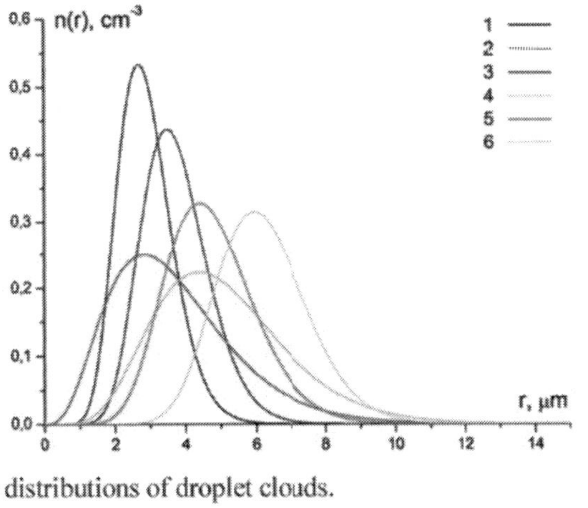

Fig. 3. The particle size distributions of droplet clouds.

The calculation was carried out for following parameters: the cloud base height is 1000 m; the penetration depth of sounding pulse is 100 m; wavelength of radiation is 532 nm.

As shown by numerical simulation results (Fig. 4), the ellipticity ε of the detected radiation of double scattering is sensitive to the microstructure parameters of the aerosol formation. The increasing of the half-width of the particle size distribution leads to the ellipticity ε varies in the smaller range for the areas of prevailing depolarization. The azimuthal dependence of the ellipticity ε are the same for models 2 and 4, and also 5 and 6. It allows us to suggest the change of the modal radius of particles doesn't influence on the ellipticity ε.

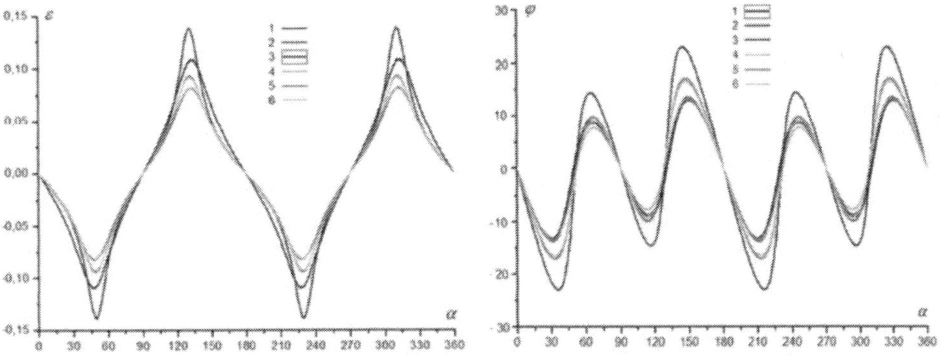

Fig. 4. Azimuthal distribution of a) the ellipticity ε and b) the semi-major axis orientation φ of the polarization ellipse of the double scattering lidar return over the receiver focal plane.

The azimuthal dependence of the semi-major axis orientation φ of the polarization ellipse of the double scattering radiation is shown in Figure 4b. Obvious influence of the droplet cloud microstructure on the orientation of the polarization ellipse is not observed. At the same time clouds with similar values of the half-width of the particle size distribution demonstrate almost complete coincidence of the azimuthal dependence of the orientation of the polarization ellipse.

It's worth focusing on the droplet cloud model 3 (tab.2). The azimuthal dependence of considered polarization parameters of the double scattering radiation from this model shows the greatest deviation from zero. It indicates that this cloud model transforms the linear polarized sounding radiation to the elliptical polarized radiation more significant at the double scattering by cloud particles. The orientation of the semi-major axis of the polarization ellipse varies over a wider range than for other cloud models. This fact needs special research.

This research was supported by "The Tomsk State University Academic D.I. Mendeleev Fund Program" grant № 8.1.12.2015.

References

1. E.D. Hinkley, *Laser monitoring of the atmosphere* (New York, 1976)
2. M.R. Measures, *Laser remote sensing* (New York, 1984)
3. V.A. Donchenko, I.V. Samokhvalov, G.G. Matvienko, Atmospheric and oceanic physics, 7,11 (1971)
4. G.M. Krekov, S.I. Kavkianov, M.M. Krekova, *Interpretation of the optical atmospheric sounding signals* (Novosibirsk, 1987)
5. R. J. Allen and C. M. R. Plaft, Appl. Opt., 16, 12 (1977)
6. D.L. Hult, L.R. Bissonnette, L. Durand Appl. Opt., 33, 12 (1994)
7. V.V. Bryukhanova, I.V. Samokhvalov, A.I. Abramochkin, S.A. Abramochkin, A.A. Tikhomirov, Atmospheric and oceanic optics 16, 09 (2003)
8. L.I. Chaikovskaya, E.P. Zege, I.L. Katsev et al., Appl. Opt., 48, 623–632 (2009)
9. V.V. Bryukhanova, I.V. Samokhvalov, G.V. Simonova, Patent 1006336 Russian Federation, IPC G01S 17/88. The device registration lidar signal / the applicant and the patentee National Research Tomsk State University and the Institute of Monitoring of Climatic and Ecological Systems, Siberian Branch of the Russian Academy of Sciences. – published 20.12.10, bulletin № 35 . – 2 p.
10. B.V., Kaul, I.V., Samokhvalov, Optics atmosphere and ocean, 12, 18 (2005)
11. V.V. Bryukhanova, Proc. SPIE 9292, 20th International Symposium on Atmospheric and Ocean Optics: Atmospheric Physics, 92922C-3 (2014)
12. G.V.Rozenberg, Twilight (Moscow, 1963)
13. S.N. Volkov, I.V. Samokhvalov, D. Kim, JQSRT 123, 169-175 (2013).
14. A.A. Kokhanovsky, *Cloud Optics* (Springer, 2006)
15. Kamchatka Branch FGBI "AVIAMETTELEKOM ROSHYDROMET", *Atlas of Clouds* (2015)
16. Program PolyMie to calculate optical properties (phase function or matrix, extinction and absorption coefficients) of water droplet clouds, INTAS project 01-0239, May 2005, http://osmf.sscc.ru/~smp/INTAS_01-0239/main.html (17 may 2012).
17. Nee Y.V., V.V. Bryukhanova, A.A. Doroshkevich, News of Higher Schools. Physics., 58, 8/2 (2015)
18. I.V. Samokhvalov, V.V. Bryukhanova, A.A. Doroshkevich and I.V. Zhivotenyuk, Rep. High School. Phisics, 55, 9/2 (2012)
19. A.A. Doroshkevich, V.V. Bryukhanova, I.V. Samokhvalov and A.P. Stykon, Proc. SPIE., 9292, 92922E (2014)

E3S Web of Conferences **11**, 00017 (2016) DOI: 10.1051/e3sconf/20161100017
Solar-Terrestrial Relations and Physics of Earthquakes Precursors

On the Painleve property of a hydrodynamic system

Dmitry Noshchenko[1,2,*] and *Andrey* Perezhogin[1,2,**]

[1] *Institute of Cosmophysical Researches and Radio Wave Propagation FEB RAS, 684034, Kamchatsky Kray, Paratunka, Mirnaya st., 7, Russia*
[2] *Vitus Bering Kamchatka State University, 683031, Petropavlovsk-Kamchatsky, Pogranichnaya st., 4, Russia*

Abstract. We represent conditions of hydrodynamic system when it passes the Painleve test. We use Kovaleskaya-Gambie method for fourth order ordinary differential system. We obtain Lorenz-like dynamic for this system.

Introduction

The existence of magnetic fields of planets, stars and galaxy is explained by a dynamo-mechanism [1]. Mathematical aspects of dynamo effect are reduced to solutions of MHD equations. Nonlinear terms of these equations don't allow to find an analytical solutions (except private cases). Direct number simulation of 3D magnetohydrodynamics with big Reynolds number require huge computing at supercomputers [2]. So simplified MHD systems are very important to investigate for understanding main features of dynamo.

In the paper we discuss the analytical properties of a dynamical system, which is the simplest model of the dynamo. We find relation between coefficients of system in order to pass the Painleve test.

We discuss a model of process which generates the average magnetic field by a turbulence flow of viscous incompressible fluid with the alpha effect in rotating coordinate system. It is described by magnetohydrodynamics equations.

$$\begin{aligned} &\partial_t \mathbf{v} + \mathrm{R_m}(\mathbf{v}\nabla)\mathbf{v} = \mathrm{Pm}\Delta\mathbf{v} - \nabla p - \mathrm{E}^{-1}\mathrm{Pm}(\mathbf{e}_z \times \mathbf{v}) + \mathrm{rot}\mathbf{B} \times \mathbf{B}, \\ &\partial_t \mathbf{B} = \mathrm{R_m}\mathrm{rot}(\mathbf{v} \times \mathbf{B}) + \mathrm{R}_\alpha \mathrm{rot}(\alpha\mathbf{B}) + \Delta\mathbf{B}, \\ &\nabla \cdot \mathbf{v} = 0, \\ &\nabla \cdot \mathbf{B} = 0, \end{aligned} \tag{1}$$

where $\mathbf{v}$ – the average velocity, $\mathbf{B}$ – the average magnetic field, p – pressure, $\mathbf{f}$ – mass density of external forces, α – tensor of the α-effect, $\mathrm{R_m}$ – magnetic Reynolds number, E – Ekman number, Pm – Magnetic Prandtl number, R_α – amplitude of the α-effect, $\mathbf{e}_z$ – the unit vector of axis of rotation.

* e-mail: d951039@gmail.com
** e-mail: d72156@gmail.com

We admit that field axially symmetric with respect to axis $\mathbf{e}_z$. Solenoidal fields $\mathbf{v}$ and $\mathbf{B}$ give the sum of the toroidal and poloidal components.

We represent the following decomposition of the velocity and magnetic field into the sum of time-dependent amplitudes and stationary poloidal (toroidal) fields products:

$$\begin{aligned} \mathbf{v} &= x_1(t)\mathbf{v}^T(\mathbf{r}) + x_2(t)\mathbf{v}^P(\mathbf{r}), \\ \mathbf{B} &= y_1(t)\mathbf{B}^T(\mathbf{r}) + y_2(t)\mathbf{B}^P(\mathbf{r}). \end{aligned} \tag{2}$$

Substitution of decomposition (2) in (1) gives system of amplitude's equations [3]:

$$\begin{aligned} \frac{dx_1}{dt} &= \mathrm{R_m}A_{112}x_1x_2 + \mathrm{E}^{-1}\mathrm{Pm}P_{12}x_2 + F_1 + \mathrm{R_m}L_{112}y_1y_2 - \mu_1x_1, \\ \frac{dx_2}{dt} &= \mathrm{R_m}A_{211}x_1^2 + \mathrm{E}^{-1}\mathrm{Pm}P_{21}x_1 + F_2 + \mathrm{R_m}L_{211}y_1^2 + \mathrm{R_m}L_{222}y_2^2 - \mu_2x_2, \\ \frac{dy_1}{dt} &= \mathrm{R_m}W_{112}x_1y_2 + \mathrm{R_m}W_{121}x_2y_1 + \mathrm{R}_\alpha W_{1\alpha2}y_2 - \eta_1y_1, \\ \frac{dy_2}{dt} &= \mathrm{R_m}W_{222}x_2y_2 + \mathrm{R}_\alpha W_{2\alpha1}y_1 - \eta_2y_2. \end{aligned} \tag{3}$$

In the system we take into account that in the axially symmetric case the vector lines of any poloidal field lie in planes passing through the axis of rotation, and the lines of any toroidal field perpendicular to them. Uppercase letters denote constant coefficients. They appear after application of Galerkin's method to a system. Coefficients μ_i and η_i determine the dissipation rate of velocity and magnetic fields modes from (2). In this case we assume $P_{12} = -P_{21}$ and some other relations on coefficients $A_{112} = -A_{211}$, $L_{112} = -W_{112}$, $L_{211} = -W_{121}$, $L_{222} = -W_{222}$, $\mu_1 = \mu_2$, $\eta_1 = \eta_2$, $F_1 = 0$.

System (3) is the simplest dynamo model without a kinematic effect. In this paper we investigate its analitical properties.

Simplified ODE system

We present the Painleve test on simplified ODE system. The Painleve test is neccesary condition for the Painleve property. The formal solution is given by Laurent series near a movable singularity $x - x_0$. For this aim we use Kowalevski-Gambier method [4].

Main steps of Kowalevski-Gambier method are

1. substitution $u(x) = u_0x^p$ in order to find integer value of parameter p.

2. computation coefficients $u(x) = u_jx^{p+j}$ for integer j. For every fixed j we have a linear algebraic system on coefficient u_j of Laurent series. If a system is consistent we have an uniformal solution for fixed j. For some values of j system can be overdefined. In this case coefficients of Laurent series are free. In order to determine j for which system is overdetermined we can calculate Fuchs indices.

3. If a linear algebra system of Laurent series coefficients for each Fuchs indices is consistent the ODE system passes Painleve test. It does not imply the Painleve property. If system is inconsistent ODE system passes test.

Let us discuss one simplified case of (3). We suppose that $L_{222} = L_{211} = W_{121} = W_{222} = F_1 = 0$, $F_2 = M$, $R_\alpha W_{2\alpha 1} = R_\alpha W_{1\alpha 2} = \alpha$, $P_{12} = K$, $P_{21} = -K$, $x_1 = u_1$, $x_1 = u_2$, $y_1 = u_3$, $y_2 = u_4$, independent variable $t = x$. So a reduced ODE system is

$$\begin{cases} \dot{u_1} = -\lambda u_1 + K u_2 - L u_3 u_4 \\ \dot{u_2} = -\lambda u_2 - K u_1 + M \\ \dot{u_3} = L u_1 u_4 + \alpha u_4 - u_3 \\ \dot{u_4} = \alpha u_3 - u_4 \end{cases} \tag{4}$$

where M, L, K, λ, α – independent parameters.

1 step. We substitute $u_i(x) = u_{i,0}x^{p_i}, i = 1..4$ at system (4). Main terms give next system on p_i: $p_1 - 1 = p_3 + p_4, p_2 - 1 = p_1, p_3 - 1 = p_1 + p_4, p_4 - 1 = p_3$. Solution of the linear system is $p_1 = -2, p_2 = -1, p_3 = -2, p_4 = -1$.

We find coefficients $u_{i,0}$ from system of main terms for calculated values of p_i, $i = 1..4$:

$p_1 x_1 = -L x_3 x_4, p_2 x_2 = -K x_1, p_3 x_3 = L x_1 x_4, p_4 x_4 = \alpha x_3$

From this system we get one trivial solution and four nontrivial

$$u_{1,0} = \frac{2}{L\alpha}, u_{2,0} = \frac{2K}{L\alpha}, u_{3,0} = \pm\frac{2I}{L\alpha}, u_{4,0} = \pm\frac{2I}{L},$$

where I – the imagery unit.

Let us check the first solution $u_{3,0} = \frac{2I}{L\alpha}$, $u_{4,0} = \frac{2I}{L}$, when $u_{3,0}$, $u_{4,0}$ have the same sign. In this case Fuchs indices are irrational $0, 1, 5/2 - 1/2\sqrt{17}, 5/2 + 1/2\sqrt{17}$. Because of the irrational value we need to take $u_{3,0}$, $u_{4,0}$ with the different sign. For positive values of $u_{3,0}$ and $u_{4,0}$ the Fuchs indices are -1, 1, 2, 4. For each Fuchs index we get 3 invariants:

$$Q_1 = \frac{8(\lambda - 1)}{\alpha L}, \ Q_2 = -2\frac{K^2 + \lambda^2 + 4\lambda - 4}{L\alpha}$$

$$Q_4 = ((\lambda - 2) u_{2,1} - u_{2,2}) K + 4\frac{(\lambda - 1)\left(\alpha^2\lambda - \lambda^3 - 2\lambda + 2\right)}{L\alpha} + \frac{\lambda(3\lambda - 2)(-u_{2,1}\lambda + M - u_{2,2})}{K}$$

Assuming $Q_1 = Q_2 = Q_4 = 0$ we get next set of system's parameters $\lambda = 1$, $K = \pm I$, $M = 0$.

General ODE system

In this case we assume $P_{12} = -P_{21}$ and some other relations on coefficients $A_{112} = -A_{211} = A$, $L_{112} = -W_{112} = -L$, $L_{211} = -W_{121} = -P$, $L_{222} = -W_{222} = -Q$, $\mu_1 = \mu_2 = \eta_1 = \eta_2 = 1$, $F_1 = 0$. For such coefficients system is more general than previos one.

$$\begin{aligned} \dot{u_1} &= A u_1 u_2 + K u_2 - L u_3 u_4 - u_1 \\ \dot{u_2} &= -A u_1^2 - K u_1 - P u_3^2 - Q u_4^2 + M - u_2 \\ \dot{u_3} &= L u_1 u_4 + P u_2 u_3 + \alpha u_4 - u_3 \\ \dot{u_4} &= Q u_2 u_3 + \alpha u_3 - u_4 \end{aligned} \tag{5}$$

Let us consider the Painleve test of system (5) with $A = P$. First we check if the pivot terms satisfy following equations

$p_1 - 1 = p_1 + p_2, p_1 - 1 = p_3 + p_4, p_2 - 1 = 2p_1, p_2 - 1 = 2p_3, p_2 - 1 = 2p_4,$

$p_3 - 1 = p_1 + p_4, p_3 - 1 = p_2 + p_3, p_4 - 1 = p_2 + p_3$

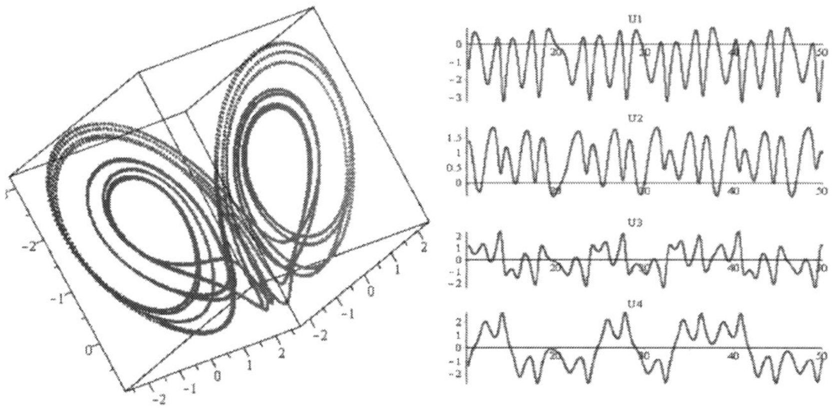

Figure 1. One velocity and two magnetic cmponents

There is one case when $p_1 = -1, p_2 = -1, p_3 = -1, p_4 = -1$. Now we substitute expantions $u_i = u_{0,i}x^{p_i}$ into pivot terms and get nonlinear system for coefficients $u_{0,i}$.

$u_{0,1} = Iu_{0,3}, u_{0,2} = (LQu_{0,1} - P)^{-1}, u_{0,3} = \frac{IL+\sqrt{-L^2-4PQ}}{2Q^2}, u_{0,4} = -Qu_{0,2}u_{0,3}$

In order to simplify $u_{0,i}$ we put $Q = -L^2/(4P)$. Such simplification gives a very simple characteristic equation for the Fuchs indices $j\left(j^3 - 6j^2 + 5j + 12\right)$. Solving equation we obtain indicies $-1, 0, 3, 4$. A characteristic polynomial doesn't depend on parameters L, P. There are two invariants for this system. Invariants correspond to the Fuchs values $j = 3$ and $j = 4$. Detailed analysis of the invariants gives some relations on coefficients. So $L = 2P$, $P = 1$. Free parameters are K, α and dependent one is M. Resulting system passes the Painleve test with $\alpha = 3$, $K = 2$.

$$\begin{aligned}
\dot{u_1} &= u_1u_2 - u_1 + 2u_2 - 2u_3u_4, \\
\dot{u_2} &= -u_1^2 - u_2 - 2u_1 - u_3^2 + u_4^2 + \tfrac{1330}{99}, \\
\dot{u_3} &= 2u_1u_4 + u_2u_3 + 3u_4 - u_3, \\
\dot{u_4} &= -u_2u_3 + 3u_3 - u_4
\end{aligned} \tag{6}$$

If we put value of M to zero, we get a modification of (6). Direct solutions of this system gives periodic oscillations (fig. 1). If we keep constant force M, the periodic solutions will be relaxation oscillations.

Conclusion

We have considered two cases of hydrodynamic system. For these systems we find sets of parameters to satisfy the Painleve test. It is possible to check what conservation law holds for this system.

References

[1] Zel'dovich B., Ruzmaikin A.A., Sokolov D.D. Magnetic fields in astrophysics. - Moscow-Izhevsk: RHD, 2006. - 384 p. In Russian.

[2] Frick P.G. Turbulence: approaches and models. - Moscow-Izhevsk: RHD, 2010. - 332 p. In Russian.

[3] Gledzer E.B., Dolzhansky F.V., Obukhov A.M. Systems of hydrodynamic type and their application. - M .: Nauka. 2010. In Russian.
[4] Conte R.M., Musette M. The Painlevé Handbook. Springer Netherlands. 2008.

On a credit oscillatory system with the inclusion of stick-slip

Roman Parovik[1,2,⋆]

[1] *Institute of Cosmophysical Research and Radio Wave Propagation of the Far Eastern Branch of Russian Academy of Science, 684034, Kamchatskiy kray, Paratunka, Mirnaya str. 7, Russia.*
[2] *Vitus Bering Kamchatka State University, 683032, Kamchatskiy kray, Petropavlovsk-Kamchatskiy, Pogranichaya str. 4, Russia.*

Abstract. The work was a mathematical model that describes the effect of the sliding attachment (stick-slip), taking into account hereditarity. This model can be regarded as a mechanical model of earthquake preparation. For such a model was proposed explicit finite- difference scheme, on which were built the waveform and phase trajectories hereditarity effect of stick-slip.

1 Introduction

The effect of stick-slip occurs into tribology problems. For example, we study movement of cargo on a spring along the surface. Due to the adhesion of the goods adheres to the surface, and due to the spring tension breaks off and slides along it at the same place his hesitation [1–3]. Also stick-slip effect may be laid in the foundation of the mechanical model of the earthquake in the subduction zone of lithospheric plates [4].

In this paper we will explore hereditarity nonlinear oscillator which characterizes the stick-slip effect [5, 6]. Recall that in [5, 6], we considered the model equation stick-slip effect, c considering hereditarity with derivatives of fractional order permanent.

In this paper we consider the oscillator hereditarity stick-slip effect with fractional derivatives variables governmental orders, which are bounded functions of. And the orders of fractional derivatives may depend not only on the transfer variable t, and shift function $x(t)$. Introduction thus fractional control parameters in the model equation generalizes considered into the third chapter hereditarity nonlinear oscillator based on stick-slip effects that can lead to new properties or effects as a result of mathematical modeling.

Next, we construct an explicit finite-difference scheme for the numerical calculation of the solution of the generalized model of stick-slip effect with variable fractional order. Next, look at some examples of the application of this scheme with different values of the control parameters, we plot and investigate phase trajectories.

2 Statement of the problem and methods of solution

A problem. Find the solution $x(t)$, where $t \in [0, T]$ the following Cauchy problem:

⋆e-mail: parovik@ikir.ru

$$\partial_{0t}^{\beta(x(t),t)}x(\tau)+\lambda\partial_{0t}^{\gamma(x(t),t)}x(\tau)+\omega^{\beta(x(t),t)}x(t)=f(x(t),t), \quad (1)$$

$$x(0)=x_0,\ \dot{x}(0)=y_0, \quad (2)$$

Where $\partial_{0t}^{\beta(x(t),t)}x(\tau)$ and $\partial_{0t}^{\gamma(x(t),t)}x(\tau)$ – the derivatives operators Gerasimova-Caputo different orders $1<\beta(x(t),t)<2, 0<\gamma(x(t),t)<1, f(x(t),t)=bt+c\sum_{n=1}^{\infty}a_n\sin(nx(t))$, b – the speed of movement of the spring, c – the energy of adhesion, λ – coefficient of friction, ω – the frequency of free oscillations, $a_n=2n\int_0^1\frac{\cos(\pi n\tau)}{\cosh^2(\pi\tau)}d\tau$ – expansion coefficients Fourier series.

Note. Gerasimov-Caputo Operators into the model equation (1) can be written as:

$$\partial_{0t}^{\beta(x(t),t)}x(\tau)=\int_0^t\frac{\ddot{x}(\tau)\,d\tau}{\Gamma(2-\beta(x(\tau),\tau))(t-\tau)^{\beta(x(\tau),\tau)-1}}, \quad (3)$$

$$\partial_{0t}^{\gamma(x(t),t)}x(\tau)=\int_0^t\frac{\dot{x}(\tau)\,d\tau}{\Gamma(1-\gamma(x(\tau),\tau))(t-\tau)^{\gamma(x(\tau),\tau)}}.$$

Therefore, considering the representations (3) are placed two Cauchy problems (1) and (2).

Note. The Cauchy problem (1) and (2) when the fractional parameters β and γ are constants, becomes the Cauchy problem in [5], if these constants are $\beta=2$ and $\gamma=1$, we arrive at the Cauchy problem for the classical stick-slip effect [3].

The Cauchy problem (1) and (2) generally has no clear decision, so we will solve this problem numerically using the theory of finite difference schemes [7, 8].

We construct an explicit finite-difference scheme, because it is easiest to implement on a computer. We divide the time interval $[0,T]$ for N equal parts increments $\tau=T/N$. Then the function solutions $x(t)$ of the differential of the Cauchy problem (1) and (2) pass grid function in $x(t_j)$, where $t_j=j\tau$, $j=0,1,\cdots,N-1$, which will be solution of the corresponding Cauchy difference problem.

Let us more detail on the approximation of the operators of fractional differentiation of (3). Following definitions [9], we obtain the following for their discrete counterparts operators (3):

$$\partial_{0t}^{\beta(x(t),t)}x(\tau)\approx\sum_{k=0}^{j-1}\frac{p_k^j\tau^{-\beta(x_k,t_k)}}{\Gamma(3-\beta(x_k,t_k))}\left(x_{j-k-1}-2x_{j-k}+x_{j-k+1}\right), \quad (4)$$

$$\partial_{0t}^{\gamma(x(t),t)}x(\tau)\approx\sum_{k=0}^{j-1}\frac{q_k^j\tau^{-\gamma(x_k,t_k)}}{\Gamma(2-\gamma(x_k,t_k))}j\left(x_{j-k-1}-2x_{j-k}+x_{j-k+1}\right), \quad (5)$$

$$p_j^k=(k+1)^{2-\beta(x_j,t_j)}-k^{2-\beta(x_j,t_j)},\ p_j^k=(k+1)^{1-\gamma(x_j,t_j)}-k^{1-\gamma(x_j,t_j)}.$$

Note. In (4)-(5) was used: $x(t_j)=x_j$.

Subject to the initial conditions (2), put relations (4)-(5) into equation (1) we get explicit finite-difference scheme:

$$x_1=\tau y_0+x_0, j=0,$$

$$x_{j+1}=Ax_j-Bx_{j-1}-D\sum_{k=1}^{j-1}\frac{\tau^{-\beta(x(t_k),t_k)}}{\Gamma(3-\beta(x(t_k),t_k))}p_k^j\left(x_{j-k+1}-2x_{j-k}+x_{j-k-1}\right)- \quad (6)$$

$$-D\sum_{k=1}^{j-1}\frac{\lambda\tau^{-\gamma(x(t_k),t_k)}q_k^j\left(x_{j-k+1}-x_{j-k}\right)}{\Gamma\left(2-\gamma\left(x\left(t_k\right),t_k\right)\right)}+Df_j,\ j=1,...,N-1,$$

$$A=\frac{2A_0+B_0-\omega^{\beta(x(t_j),t_j}}{A_0+B_0},B=\frac{A_0}{A_0+B_0},\eta=\frac{\omega^{\beta(x(t_j),t_j)}}{A_0+B_0},$$

$$D=\frac{1}{A_0+B_0},A_0=\frac{\tau^{-\beta(x_0,t_0)}}{\Gamma\left(3-\beta\left(x_0,t_0\right)\right)},B_0=\frac{\lambda\tau^{-\gamma(x_0,t_0)}}{\Gamma\left(2-\gamma\left(x_0,t_0\right)\right)},$$

$$p_k^j=(k+1)^{2-\beta(x(t_k),t_k)}-k^{2-\beta(x(t_k),t_k)},q_k^j=(k+1)^{1-\gamma(x(t_k),t_k)}-k^{1-\gamma(x(t_k),t_k)}.$$

Note. Explicit finite difference scheme as are usually conditional stable, ie, there is a limitation on the step τ of difference grid. The scheme (6) are no exceptions. The step of the grid can be controlled by the method of double recalculation or Runge rule. In this paper we carry out a numerical experiment and see how the maximum absolute value error between the exact and numerical solutions, depending on the increase in the number of N nodes of the computational grid.

Note. The explicit finite difference scheme (6) in the case where the fractional parameters are constants and have values of $\beta = 2$ and $\gamma = 1$, become to the local settlement explicit finite-difference scheme for the classical optical effect stick-slip [3].

3 Results of numerical simulation

In [3] it was said that it is enough to take the first seven coefficients a_k in the expansion of the function $f(x(t), t)$ to obtain reliable solutions (1). The values of these coefficients take from [3] $a_1 = 0.436, a_2 = 0.344, a_3 = 0.164, a_4 = 0.058, a_4 = 0.021, a_6 = 0.004, a_7 = 0.003$.

Numerical modeling hereditarity stick-slip effect on the formulas (6) and obtained in [5] depending on the control parameters. First, we consider the dependence of the fractional control parameters only on the time t, ie, $\beta = \beta(t)$ and $\gamma = \gamma(t)$. Then we look at the more general case, when the fractional parameters also depend on $x(t)$ of displacement.

Example 1. The control parameters: $\beta(t) = 1.8 - 0.03\sin(\pi t), \gamma(t) = 0.6 - 0.04\cos(\pi t), N = 3000, \delta = 50, \tau = 0.05, \lambda = 0.3, b = 1, \omega = 1, y_0 = 0.3, x_0 = 0$.

In Fig.1 shows the calculated curves of displacement, velocity displacement and the phase trajectory. In Fig.1a shows an oscillogramma for example 1. It can be seen that in the separation of the load fluctuates, the rate of these oscillations in the potential well decays slowly enough (Fig.1b). This effect is ereditarnostyu process. In the phase trajectory (Fig.1c) shows that potential wells are stable focuses.

Example 2. The control parameters: $\beta(t) = 1.6 - 0.1\sin(\pi t), \gamma(t) = 0.6 - 0.2\cos(\pi t), N = 3000, \tau = 0.05, \delta = 50, \lambda = 0.3, b = 1, \omega = 1, y_0 = 0.3, x_0 = 0$.

In Fig.2 shows the calculated curves obtained by numerical formulas (6) and [5]. Unlike the examples discussed above, in this example the fractional parameters depend not only on the time t, but on the $x(t)$ displacement.

In Fig.2a shows the oscillogrammas obtained by the formulas (6) and [5], which characterize the effect of stick-slip given hereditarity. You may notice that the goods slide along the surface, testing damped oscillations of small amplitude. It can be seen that the amplitude of these oscillations increases slightly for the subsequent potential wells (Fig.2b). Phase trajectories to this case are given in Fig.2c.

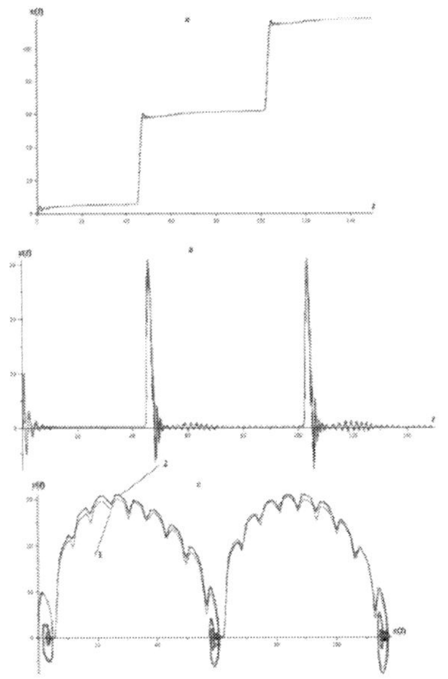

Figure 1. The calculated curves obtained from the formulas (6) (curve 1) and [5] (curve 2): a) - oscillogramma, b) - oscillator speed, c) - the phase trajectory

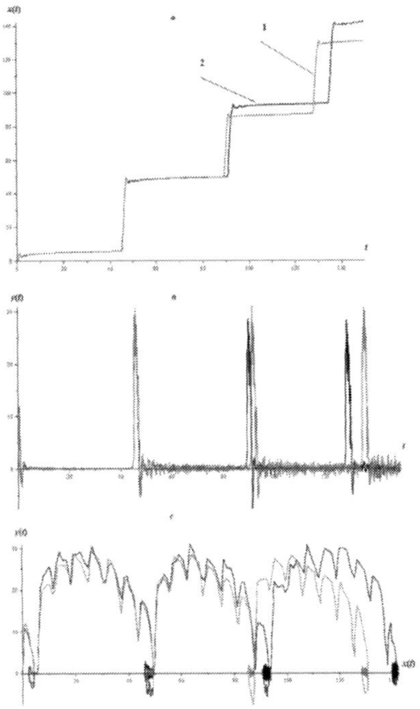

Figure 2. The calculated curves obtained from the formulas (6) and [5] a) - oscillogramma b) - oscillator speed, c) - the phase trajectory

4 Conclusion

It can be concluded that both finite-difference scheme (6) give good agreement with the results of [5]. In the future we plan a more detailed study of the proposed finite-difference.

References

[1] Uspekhi fizicheskikh nauk, Dedkov G. V., **170**:6, 585-618 (2000)
[2] Physical Review E Daub E. G., Carlson J. M., **80**:6, 066113 (2009)
[3] Rekhviashvili S. Sh., *Razmernye yavleniya v fizike kondensirovannogo sostoyaniya i nanotekhnologiyakh* (KBNTs RAN, Nal'chik,Russia, 2014) 250 p.
[4] Scholz Ch. H., *The mechanics of earthquakes and faulting* (Cambridge university press, 2002) 471 p.
[5] Doklady Adygskoy (Cherkesskoy) Mezhdunarodnoy Akademii Nauk, Parovik R.I.,**17**:3, 70-77 (2015)
[6] Parovik R.I., *Proceedings of International Russian-Chinese Conference on Actual Problems of Applied Mathematics and Problems and Modern Problems of Algebra, Analysis and Informatics* (December 14-18, Elbrus, Kabardino-Balkarian Republic, Russia, 2015) 159-160
[7] Samarskiy A. A., Gulin A. V., *Ustoychivost' raznostnykh skhem*, (Nauka, Moscow, 1973)
[8] Samarskiy A. A., *Teoriya raznostnykh skhem*, (Nauka, Moscow, 1977) 656 p.
[9] Parovik R. I., *Matematicheskoe modelirovanie lineynykh ereditarnykh ostsillyatorov* (KamGU im. Vitusa Beringa, Petropavlovsk-Kamchatskiy, Russia, 2015) 178 p.

Multiparameter monitoring of short-term earthquake precursors and its physical basis. Implementation in the Kamchatka region

Sergey Pulinets[1,*], *Dimitar* Ouzounov[2], *Dmitry* Davydenko[1] and *Alexei* Petrukhin[3]

[1]Space Research Institute, RAS, 84/32 Profsoyuznaya str., 117997, Moscow, Russia
[2]CEESMO, Chapman University, 1 University Drive, Orange, CA 92866, USA
[3] Schmidt Institute of Physics of the Earth, RAS, Bolshaya Gruzinskaya, 10, 123242, Moscow, Russia

Abstract. We apply experimental approach of the multiparameter monitoring of short-term earthquake precursors which reliability was confirmed by the Lithosphere-Atmosphere-Ionosphere Coupling (LAIC) model created recently [1]. A key element of the model is the process of Ion induced Nucleation (IIN) and formation of cluster ions occurring as a result of the ionization of near surface air layer by radon emanating from the Earth's crust within the earthquake preparation zone. This process is similar to the formation of droplet's embryos for cloud formation under action of galactic cosmic rays. The consequence of this process is the generation of a number of precursors that can be divided into two groups: a) thermal and meteorological, and b) electromagnetic and ionospheric. We demonstrate elements of prospective monitoring of some strong earthquakes in Kamchatka region and statistical results for the Chemical potential correction parameter for more than 10 years of observations for earthquakes with M≥6. As some experimental attempt, the data of Kamchatka volcanoes monitoring will be demonstrated.

1 Introduction

We consider the final stage of the earthquake preparation cycle as the process of approaching the open complexed system to a critical state, and atmosphere (and its ionized upper shell-ionosphere) are the integrated parts of this system. With this approach, it can be argued that the crustal deformation, foreshock activity, various geophysical anomalies recorded by ground-based monitoring methods, atmospheric and ionospheric anomalies registered from space are not independent processes. They are the parts of synergetic interaction of geospheres, and by their distribution in space, by the sequence of appearance in time allow to reveal what is called in synergetics "arrow of time" characterizing the irreversibility of the process. In the case when a core set of key parameters is available (as it was after the analysis of numerous publications on the earthquake in Aquila April 6, 2009) such directionality can be demonstrated experimentally (see Fig. 2 in [1]). However, the organization of operational forecasting by monitoring the large number of the necessary

* Corresponding author: pulse@rssi.ru

data is not easy task, and new approaches should be investigated. We offer the use of so-called integral parameters characteristic to a synergistic approach. The advantage of these options is their availability (remote access), global accessibility (most used parameters - a remote sensing data from satellites) and reliability (according to our experience, their abnormal behavior seen in almost 100% of cases of strong M> 6 earthquakes). These parameters are called integral because they characterize the state of the environment as a whole and reflect a significant change in its properties.

Our additional and very important task is to unite the probabilistic and deterministic approaches to the earthquake forecast what will prepare ground to physicists and seismologists to work together on this noble task.

2 Integral parameters

To demonstrate the rightness of our approach among large set of the physical precursors of earthquakes we selected two: one – the *b*- value from the Gutenberg-Richter Frequency-Magnitude Relationship (FMR), and second – the correction of atmospheric chemical potential (ACP) which indicated the deviation of the latent heat transform in the boundary layer of atmosphere in the presence of ionization source. The first parameter is used in probabilistic approach by seismologists, and the second one – in deterministic approach exploring the physical short-term precursors.

2.1 *b*-value as indicator of precursory period

To characterise the regional seismicity Gutenberg and Richter [2] proposed the empirical relationship in the form:

$$\log N = a - bM \quad (1)$$

where N – is a cumulative number of earthquakes with the magnitude higher than M, a – coefficient characterising the regional seismicity, and b – the parameter that seems characterises the state of the crust on different stages of the earthquake cycle. Usually, its value is closed to 1, but on different studies of the seismic cycle it varies essentially and these variations look like indicator of the main shock approaching. At least in [3] the authors claim that "lower than average *b*-values characterize locked patches of faults (asperities), from which future mainshocks are more likely to be generated". In [4] the *b*-value drop was established as one of the three parameters which are used to identify the foreshock activity. This parameter attracted our attention because the *b*-value drop coincides in time with the period of increased radon emanation. In the left panels of the Fig. 1 the illustrations from the papers [5] and [6] are combined. Upper part of the left panel from top to bottom presents the seismic activity around the time of M6.9 Kobe earthquake which took place on 17 January 1995 in Japan, *b*-value, and fractal dimension of the seismic flux approximately equal to 2*b*. The lowest graph in the left panel shows the radon concentration in water [6]. The arrow shows the time coincidence of the parameters shown in the upper and lower graphs (to demonstrated radon variations more in details). As one can see, the sharp increase of radon emanation coincides in time with the period of *b*-value drop. However, the period of the sharp increase of radon concentration one can observe the gradual increase during several months that coincides with the period of increased *b*-value.

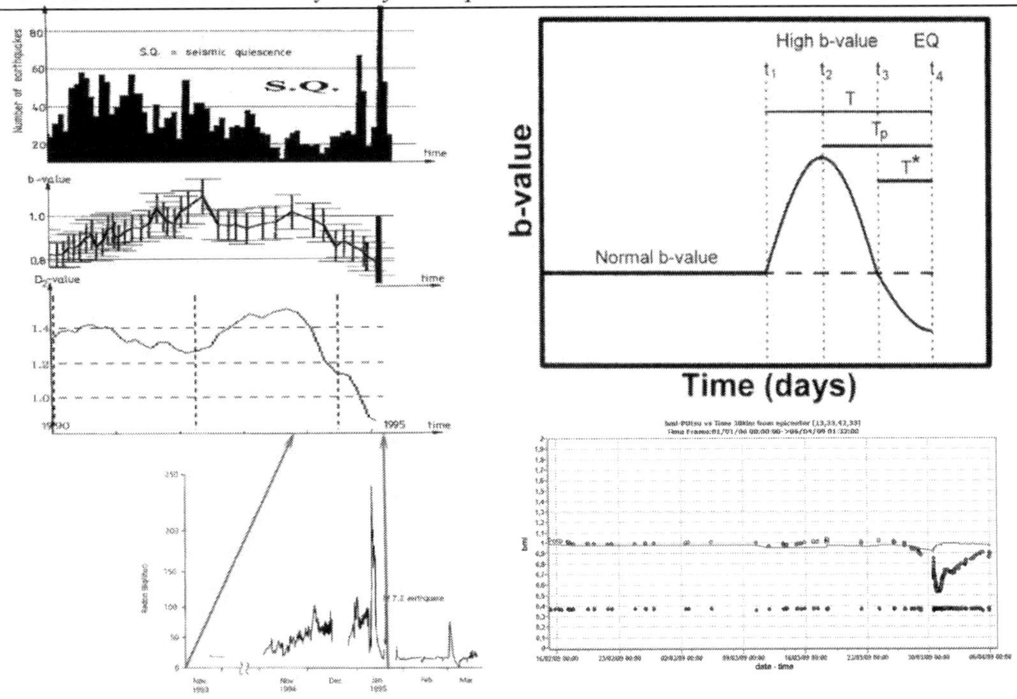

Fig. 1. Left panel from top to bottom: seismic activity around the time of Kobe earthquake, *b*-value variations, fractal dimension of the seismic activity variations after [5]), concentration of radon in water (after [6]; right panel, upper graph – schematic presentation of *b*-value anomalous variation before the main shock (after [7]), lower graph – *b*-value variations around the time of L'Aquila earthquake (after [4])

In review paper [7] the authors try to determine the precursory period with the help of *b*-value variations and determine the following phases of *b*-value anomaly before earthquake. This period denoted by T is divided by 3 phases: *b*-value increase (from t_1 to t_2), *b*-value decrease from absolute maximum to the normal value (from t_2 to t_3), and *b*-value decrease from the normal value to the anomalous minimal values (from t_3 to t_4). The period from t_2 to t_4 when continuous decrease of *b*-value observed, called the precursory period T_P. Unlike the paper [7], Papadopoulos [4] considers only the *b*-value drop lower than normal, and in his graph we can see that minimum of *b*-value is reached before the main shock. Nevertheless, we can observe the slight increase of *b*-value on his graph (right left corner of the Fig. 1). Our comparison with the time scale of anomalous variations of *b*-value and physical precursors for L'Aquila earthquake [1], show the coincidence with Papadopoulos results.

The integral character of *b*-value precursors is determined by two factors: a) it is calculated not in a single point but integrates the seismic activity within some area around the epicenter; b) in one parameter are reflected the complex processes of the earth's crust transformation from elastic deformation up to brittle changes. Decrease of *b*-value is equivalent to decrease of fractal dimension what can be interpreted as consolidation and clustering of seismic activity (observed experimentally) and increase of cracks formation. Just these new cracks are the opportunity for radon to find the new ways to the ground surface what leads (together with stress storing equivalent the pressure increase) to the amplification of the radon flux before earthquakes.

However, it is not so simple to monitor the radon activity over the large areas of earthquake preparation zone, spatially with remote sensing from satellites. Recently we have reached possibility to get the radon proxy using the secondary thermal effects in atmosphere produced by air ionization.

2.2 ACP as a proxy of radon variations

Due to its decay, radon is emitting the αparticles with energy 5.46 MeV. Taking into account the low ionization potential of atmospheric gases (10-20 eV), every αparticle emitted by radon can create in average $3{\cdot}10^5$ electron-ion pairs. The presence of ions in atmosphere creates possibility to the molecules of water vapor to join to these ions through the hydration process, which is different from condensation. In the process of condensation the chemical potential is equal to the Latent heat which is Q=40.683 kJ/mol or U_0= 0.422 eV per one molecule. The process of evaporation/condensation been the phase transition of the first order always takes place during the chemical potential equality. However, new formed ions due to air ionization by radon have the chemical potential different from the water molecules. Therefore, in the one-component approximation we introduce the correction to the chemical potential ΔU, which takes into account this difference, and the chemical potential in condition of external impact due to ionization can be expressed as:

$$U(t)=U_0+\Delta U{\cdot}cos^2t \tag{2}$$

where U_0 is the chemical potential for pure water, and $U(t)$ is the chemical potential where the ionization and hydration is taken into account. By cos^2t we take into account the daily changes of the solar radiation.

Then we can express the air relative humidity in the following form:

$$H(t)=\frac{\exp\left(-{}^{U(t)}\!/_{k\cdot T}\right)}{\exp\left(-{}^{U_0}\!/_{k\cdot T}\right)}=\exp\left(\frac{U_0-U(t)}{k\cdot T}\right)=\exp\left(-\frac{0.032\cdot\Delta U\cdot\cos^2 t}{(k\cdot T)^2}\right) \tag{3}$$

where k – Boltzman constant, and T – air temperature. The increase of the water molecules chemical potential ΔU (which can be derived from (3)) indicates the strength of nucleation process and can be used as indicator of radon activity because the nucleation is proportional to the new ions' concentration which in turn is proportional to radon activity.

No one model of atmosphere takes into account the ionization impact, therefore the variations of the air humidity and temperature resulting from the radon activity will be interpreted by models as anomalous.

The experimental check made evident that the temporal morphology of ΔU (denoted by us as ACP) is very similar to radon behavior before earthquakes, and this parameter can be used to monitor activity from space.

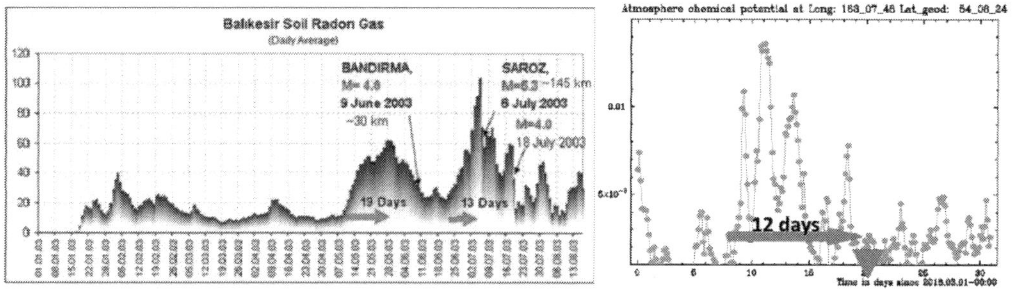

Fig. 2. Left panel – typical radon variations around the time of series of earthquake in Turkey; right panel ACP variations around the time of M6.7 earthquake on 20 March 2016 at Kamchatka

As one can see from the Fig. 2 the ACP variations practically are the same as radon variations, even the temporal scale (time from start of parameter increase to the mainshock) is identical.

3 Real time short-term precursors monitoring of Kamchatka

Multiparameter monitoring permits to increase the reliability of earthquake forecast from the point of view of the forecast parameters determination. Selection of precursors is limited by their availability for different regions of the globe. In this regard, the satellite monitoring has advantage before the ground-based measurements. For Kamchatka region the three most reliable precursors were selected: OLR (Outgoing Longwave infrared Radiation), measured at the level of top of atmosphere (10-15 km of altitude), ACP derived from the satellite measurements of the air temperature and humidity at the altitude 100 m, and GPS TEC anomalies. One can find in [1] their physical explanation and interrelation, here we want to demonstrate how the main forecast parameters are extracted from the measurements.

One of the advantages of mutiparameter monitoring is the different leading time to mainshock moment. The OLR appears the first and serves as the warning to start the real-time detailed monitoring of all parameters. Then the ACP parameter is used which has it main maximum within the time interval 18-4 days before the mainshock. And as the last usually appears the ionospheric anomaly which could be monitored both by GPS TEC or ground based vertical sounding monitoring. It is important to note that we are able using the satellite technologies to monitor not only temporal behaviour of precursory anomaly but also its spatial distribution. The left panel demonstrates the spatial distribution of OLR anomaly registered 01 of March 2016 19 days before the M6.4 earthquake on 20 March 2016 also mentioned in comments to the Fig. 2. The anomaly determines the region of the future earthquake epicenter position, while anomaly intensity permits to determine the earthquake magnitude. The right panel of the Fig. 3 shows the spatial distribution of ACP for the same earthquake registered 10 days before the earthquake. Taking into account that ACP repeats not only temporal but also the spatial behaviour of radon, from the size of anomaly we also can estimate earthquake magnitude using the Dobrovolsky formula for radius of earthquake preparation zone.

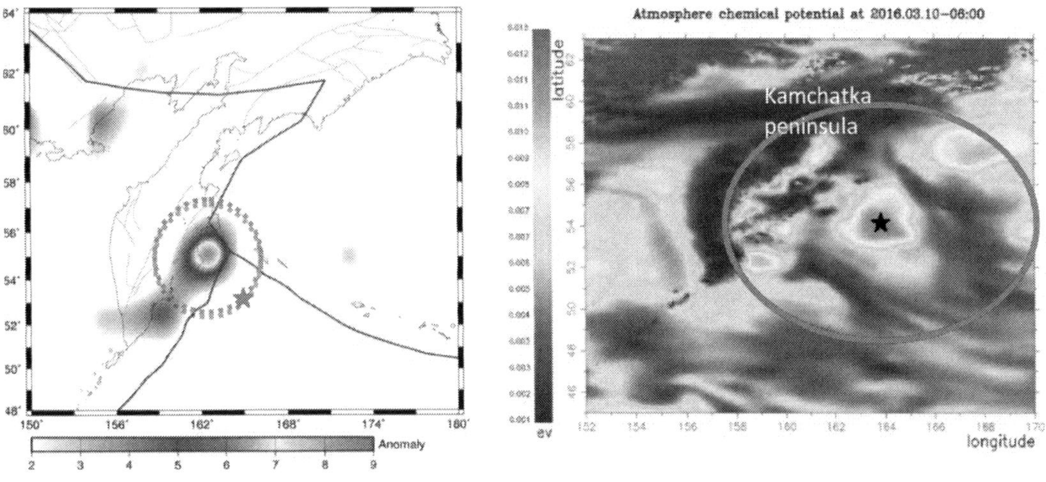

Fig. 3. Left panel – OLR anomaly registered on 01 of March 2016; right panel – ACP anomaly registered 10 of March 2016. Asterisks-epicenter position, red circle –Dobrovolsky zone

3.1 Statistical analysis of ACP variations at Kamchatka region

We analyzed variations of ACP for 64 earthquakes at Kamchatka region during more than 11 years (from 2005 till part of 2016) and revealed self-similarity of the ACP precursors through the whole territory including Kuril islands and ocean around Kamchatka. The manifestation of precursors in ocean as it is seen at the Fig 3 confirms the gaseous (radon) source of the anomaly.

The results of this analysis are demonstrated in the Fig. 4 in the form of distribution of the leading time of main maximum in time series of ACP and minimum just before the earthquake. One can see that the distribution of maxima (red) has two peaks (-11 and -4) before earthquake, while distribution for minima (blue) has major peak at the earthquake day and smaller one 2 days before earthquake

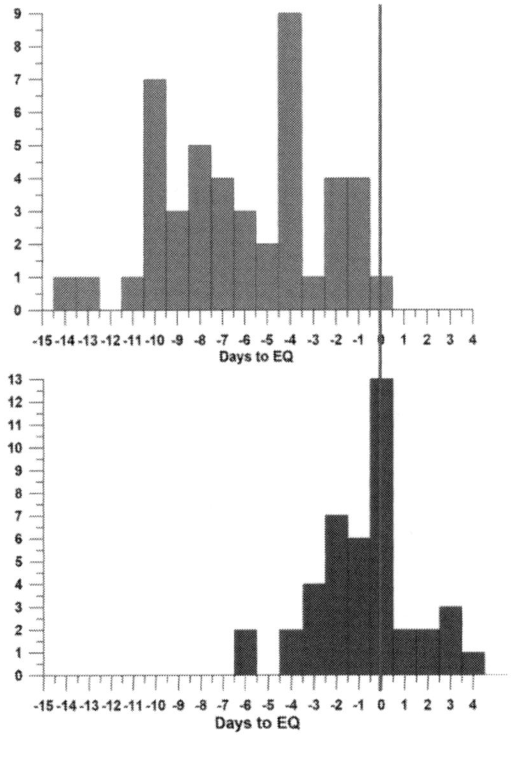

Fig. 4. Upper panel – distribution of leading time of the main maxima of the ACP time series; lower panel – distribution of leading time of the main miniima of the ACP time series. Vertical red line indicates the day of earthquake.

3.2 Ionospheric monitoring of Kamchatka region

Unfortunately we had limited access to the data of Kamchatka GPS receivers network. Practically the data of only two receivers of IGS network were available (*pets* and *yssk*) what permitted us to provide the so called cross-correlation analysis of the ionospheric anomalies before earthquake at Kamchatka [8]. Advantage of this technique is in its imminent forecast ability: the drop of the cross-correlation coefficient is observed usually one day before the main shock. In the Fig. 5 is shown the example of real time monitoring of the GPS TEC for period of June-July 2012 when the series of M≥6 earthquake took place in the region. The effectiveness of the technology is obvious.

DOI: 10.1051/e3sconf/20161100019

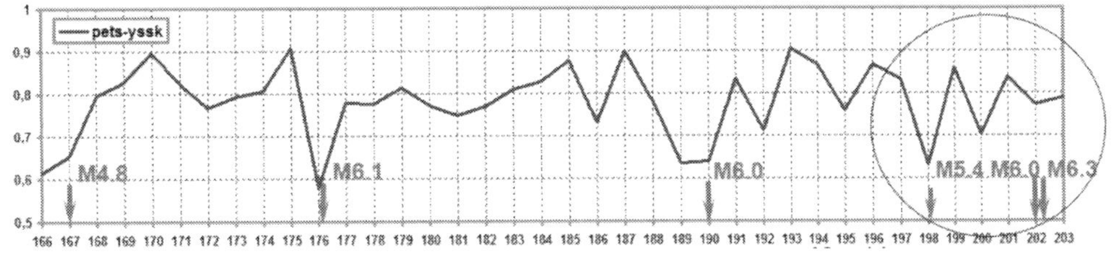

Fig. 5. Cross-correlation coefficient calculated for period June-July 2012 for GPS receivers *pets* and *yssk*.

4 Volcanoes monitoring by ACP technique

Existing reports on the radon activity during the volcanoes eruption [9] inspired us to check how the ACP variations correspond to the reported radon variations during volcano eruption. The results of comparison appeared amazing: the complete coincidence of radon and ACP variations was revealed. In the left panel of Fig. 6 the radon variations during Taal volcano (Philippines) eruption in October1994 are shown. In the right panel the Klyuchevskoi volcano activity in ACP for period from 1 August til 16 September 2016 is shown.

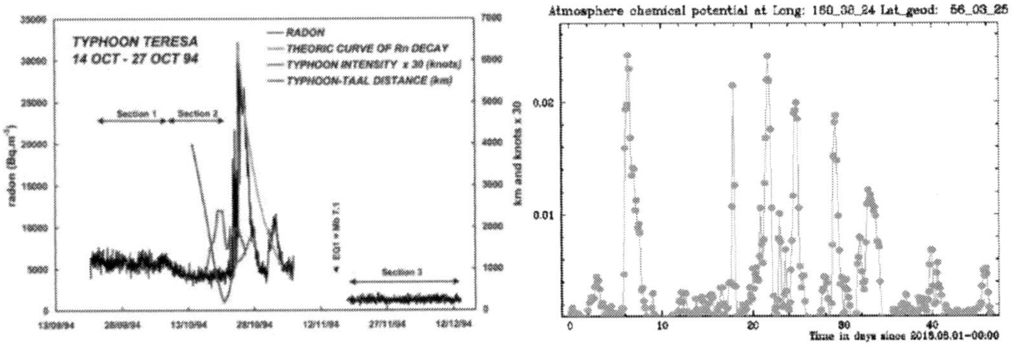

Fig. 6. Left panel – radon activity (black line) and radon decay curve (green line) are shown during Taal volcan eruption on October 1994; right panel – variations of ACP during Klyuchevskoi volcano activity in August-September 2016.

References

1. S.A. Pulinets, D.P. Ouzounov, A.V. Karelin, D.V. Davydenko, Geomagn. Aeron., **55**, 540 (2015)
2. B. Gutenberg, C.F. Richter, Bull. Seism. Soc. Am., **34**, 185 (1944)
3. D. Schorlemmer, S. Wiemer, M. Wyss, JGR, **109**, *B12307*, (2004)
4. G.A. Papadopoulos, M. Charalampakis, A. Fokaefs, G. Minadakis, Nat. Haz. Earth Syst. Sci., **10**, 19 (2010)
5. B. Enescu, K. Itu, Annuals of Disast. Prev. Res. Inst., Kyoto Univ., No.46 B (2003)
6. G. Igarashi, S. Saeki, N. Takahata, and K. Sumikawa, Science, **269**, 60 (2010)
7. J.-H. Wang, K.-C. Chen, P.-L. Leu, C.-H. Chang, J. Seismology, **20**, 905, (2016)
8. S.A. Pulinets, T.B. Gaivoronska, A. Leyva Contreras, L. Ciraolo, , Nat. Haz. Earth Syst. Sci., **4**, 697 (2004)
9. P. Richon, J.-C. Sabroux, M. Halbwachs, et al., GRL, **30**, 1481 (2003)

Joint perturbation of geoacoustic, emanation, and atmospheric electric fields at the boundary of the earth's crust and the atmosphere before an earthquake

Oleg Rulenko[1,2,*], *Yuriy* Marapulets[2,**], *Yuriy* Kuzmin[3,***], and *Alexandra* Solodchuk[2,****]

[1] *Institute of Volcanology and Seismology, Far Eastern Branch, Russian Academy of Sciences, Russia, 683006, Petropavlovsk-Kamchatskiy, boulevard Piypa, 9*
[2] *Institute of Cosmophysical Research ans Radio Wave Propagation, Far Eastern Branch, Russian Academy of Sciences, Russia, 684034, Kamchatska region, Elizovskiy district, Paratunka, Mirnaya str., 7*
[3] *Kamchatka Branch, Geophysical Service, Russian Academy of Sciences, Russia, 683006, Petropavlovsk-Kamchatskiy, boulevard Piypa, 9*

Abstract. The possibility of joint perturbation of geoacoustic, emanation and atmospheric electric fields by the Earth's crust — atmosphere boundary before an earthquake is considered. In order to detect it, simultaneous registration of these fields was carried out in Kamchatka in August–October 2012. Synchronous anomalous disturbances were recorded 13 days before an earthquake with the magnitude $M_w = 5.6$ at the distance of 140 km from an epicenter. It may be considered to be a complex short-term precursor.

1 Introduction

The boundary between the lithosphere and the atmosphere is characterized by significant changes of properties of the contacting mediums, considerable flows of mass and energy, a wide list of simultaneously existing dissimilar fields and their interactions and transformations [1]. A constituent part of interaction of the solid and the gas geospheres is the lithosphere effect on the atmosphere. It manifests partially in anomalous changes of geophysical fields in the near ground atmosphere in the result of geodynamic processes in the Earth's crust and is intensive in seismically active regions at the final stage of earthquake preparation.

Among different near-surface fields, the joint behavior of geoacoustic, emanation, and atmospheric electric fields before earthquakes deserves our attention. By the present time, many researchers have determined pre-seismic perturbations of these different fields, but they have not been registered simultaneously. Joint analysis of the perturbations of different fields improves the reliability of detection of earthquake precursors and may be the basis for investigation of the process of their preparation [2], [3].

*e-mail: rulenko@kscnet.ru
**e-mail: marpl@ikir.ru
***e-mail: kuzy@emsd.ru
****e-mail: aleksandra@ikir.ru

It is generally accepted that mechanic processes play the main role in the preparation of a tectonic earthquake. They cause changes of the Earth's crust stress state which in the result of its inhomogeneous hierarchical block structure determine complicated spatial-temporal deformations of rocks, in particular their local tensions and compressions. Fig. 1 shows a scheme of generation of joint perturbation of geoacoustic, emanation and atmospheric field by the boundary of the Earth's crust and the atmosphere before an earthquake in the presence of sedimentary rocks and their tension at the observation site. The scheme was made by the authors on the basis of colligation and analysis of the results of theoretical investigations and natural observations from the world literature. Pre-seismic perturbations of these fields just like precursors in other fields, cause changes of the stress field in the zone of earthquake preparation. They occur in the result of transformations of stress-deformation tensions of the near-surface sedimentary rocks at an observation point. They have their peculiarities of genesis but common deformation nature.

Tension of sedimentary rocks having polydisperse fluid-saturated porous low-tensile structure is accompanied by relative displacement of fragments, their rehandling and, as a consequence, generation of acoustic signals. Simultaneously with that, in the result of opening of contacting surfaces, increase of rock emanation, opening of isolated pores and crack widening, radon and thoron concentrations in the soil gas increase as well as their penetration into the atmosphere that increases near-surface air ionization. In case of fair weather conditions, this process is accompanied by generation of negative electric charge in the near-surface layer of the thickness from tens of meter fractions to several meters [4], [5]. Its density may reach high values. Near such a charge and lower, the normal atmospheric electric field decreases and even changes its sign ([6] — short review). Generation and dissipation of the negative volume charge causes electric field bay decrease which is observed before earthquakes [7]–[9]. The general factor promoting the considered processes is the fragmentariness of sedimentary rocks and, as a consequence, their high deformability.

2 Observations and results

In August–October 2012, volumetric activity of radon Rn and thoron Tn in the surface layer, geoacoustic emission in the frequency range of 0.7–2.0 kHz, atmospheric electric field potential gradient at the altitude of 7 cm from the ground surface, air pressure, wind velocity and rain intensity were simultaneously measured for the first time at "Karymshina" site (52.83°N, 158.13°E) of the Institute of Cosmophysical Research and Radio Wave Propagation FEB RAS [6]. Stored energy of earthquakes occurred at the distance of $R \leq 250$ km from this site was considered to be the characteristic of the local seismic activity. Data of the regional catalogue of Kamchatka Branch of Geophysical Service RAS were applied (http://emsd.ru).

Fig. 2 illustrates a graph of energy storage of the considered earthquakes, air pressure variations and emanation volumetric activity. It is clear from Fig. 2, that high increases of Rn and Tn were only on October 2. They lasted for about 8 hours. Increase of emanation volumetric activity, electric field decrease and geoacoutsic emission increase at three measurement points forming a triangle with the sides of 10–35 m occurred during fair weather conditions (absence of rain, strong or moderate wind, air pressure was weakly changing, Fig. 3). During the period of measurements of 52 days, geoacoustic emission increases and close to it in time the electric field decreases with sing change were not observed any more. That indicates non-meteorological nature of these perturbations.

Joint anomalous perturbation of the fields was observed 13 days before the earthquake with $K_s = 13.5$ which occurred on October 15 2012 at 01:19 UTC at the distance of 140 km from "Karymshina" site. Epicenter coordinates are 51.900°N, 159.379°E, the depth of a source is 21 km, moment magnitude $M_w = 5.6$ (NEIC catalogue, http://earthquake.usgs.gov). This earthquake was accompanied by aftershocks. The strongest occurred on October 15 at 22:45. It had $K_s = 12.6$, the

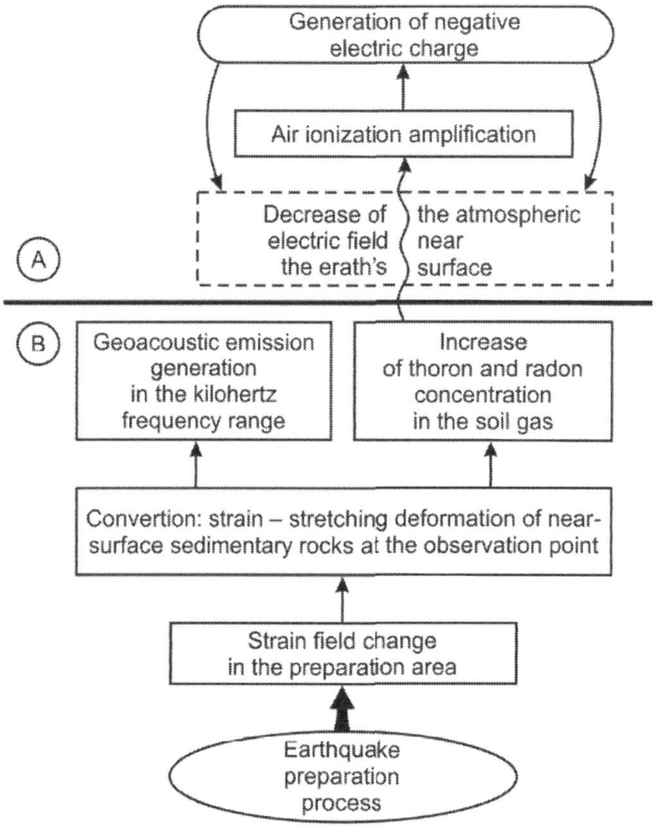

Figure 1. Scheme of generation of the joint perturbation of geoacoustic, emanation and atmospheric electric fields by the boundary of the Earth's crust and atmosphere before an earthquake. A — ground air, B — near-surface sedimentary rocks.

epicenter coordinates are 51.810°N, 159.519°E, the depth of a source is 35 km. It is clearly seen on the graph of seismic energy storage (Fig. 2).

A well No. 99-8 of Kamchatka Branch of Geophysical Service RAS is located 120 m from the measurement field. Its depth is 19 m. Its lithologic log is presented by the following rocks: wash-pebbly deposits with sandy-argillaceous filling (0—5 m), wash-rubbly deposits with argillaceous filling (5—14 m), wash-pebbly deposits with sandy filling (14— 19 m). Water statistic level is 13.7 m. Thus, near-surface rocks in "Karymshina" site are sedimentary. The thickness of the layer is 50 m.

The theoretical radius of precursor manifestations in the Earth's surface deformations, known as $r = e^M$ (km), is 270 km for the main earthquake on October 15. It is significantly larger than the epicentral distance of "Karymshina" site which is located in the zone of intersection of tectonic faults of different ranks. Fault zones are characterized by increased tectonic sensitivity of different fields and by their stronger variations [10], as well as by stronger perturbations of fields before earthquakes [3]. Thus, the seismogenic nature of anomalous perturbations on October 2 is quite real.

Anomalous increase of radon and thoron with the duration of about 8 h 13 days before the earthquake on October 15 with $M_w = 5.6$, $R = 140$ km agrees well with the results of the paper [11]. In that

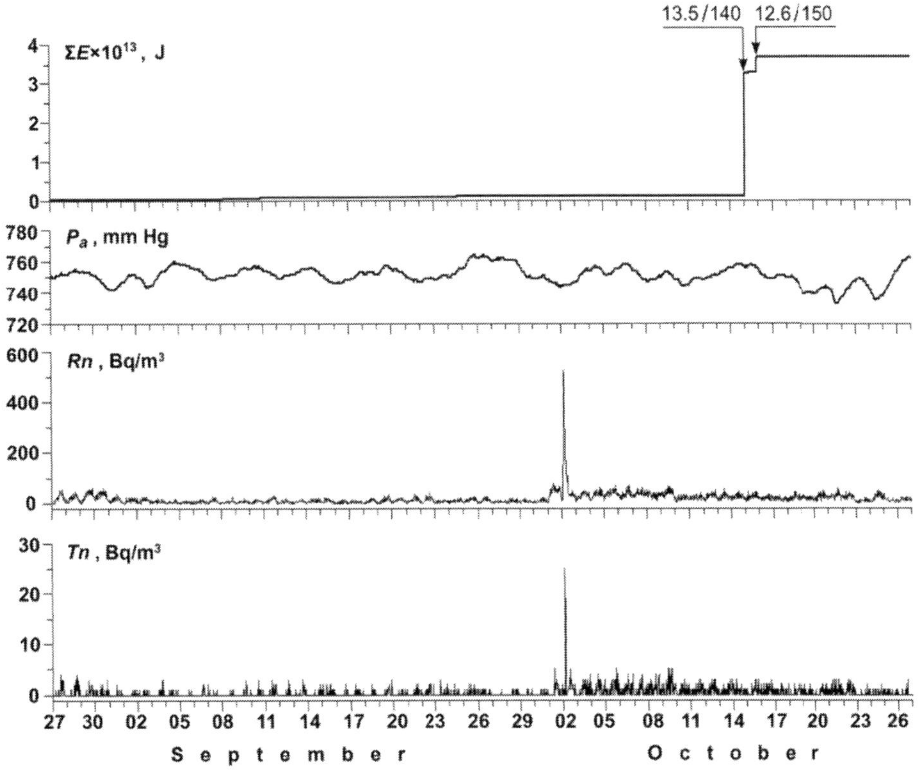

Figure 2. Graph of energy storage $\sum E$ of earthquakes with $K_s > 9.0$ which occurred in the region of sampling, air pressure variations P_a, volumetric activity of radon Rn and thoron Tn during the observations. The left arrow indicates the time of the main earthquake on October 15 2012, the right arrow indicates the strongest aftershock. The first number over the error is K_s, the second number is the epicentral distance, km.

paper, simultaneous increases of radon and toron in the soil gas before an earthquake with $M_L \geq 4.5$, $R < 150$ km was registered in the zone of an active fault in Taiwan. They usually occurred 1–20 days before an event and had the least duration of 5–7 h.

Radon and thoron differ insignificantly in physical and chemical properties except for the half-life which is 3.825 days and 55.6 s, respectively. Due to the short period of half-life, thoron cannot migrate far from the place of formation, its mass transfer is of surface character. When entering into the atmosphere, it comes out of the soil layer with the thickness of 6 cm. Its first 2 cm give up to 75% of emanation [12]. The large period of half-life allows the radon to come out from the depths of 10—12 m by diffusive mass transfer during the time equal to three–four periods [13]. During sedimentary rock tension, this depth becomes larger.

The anomalous disturbance of geoacoustic, emanation and atmospheric electric fields registered on October 2 may be considered to be a complex short-term precursor of the earthquake which occurred on October 15 with $M_w = 5.6$ and as the indication of the Earth's crust effect on the surface atmosphere during its preparation. The affecting physical agents were radon and thoron. Stronger outflows of these emanations from the ground increases surface air ionization. It is accompanied by generation of negative volume charge in fair weather conditions well known in the atmospheric electricity. It decreases the electric field.

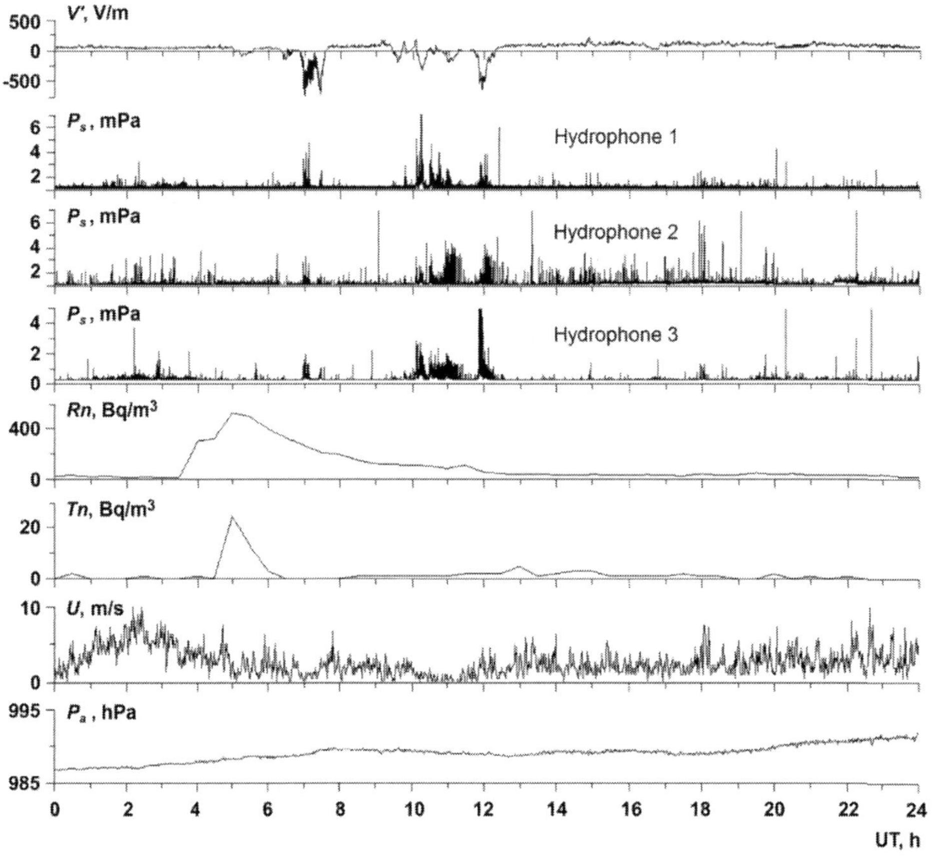

Figure 3. Variations of potential gradient V' of the atmospheric electric field, of acoustic pressure P_s at three measurement points, of volumetric activity of emanations, wind velocity U at the height of 8 m from the ground and atmospheric pressure P_a on October 2, 2012.

3 Conclusions

We have suggested a scheme of joint perturbation of geoacoustic, emanation and atmospheric electric fields by the boundary of the Earth's crust and atmosphere before the earthquake in the presence of sedimentary rocks and their tension at the observation site. Simultaneous registration of these fields has been carried out for the first time. Their joint anomalous disturbance has been discovered. It was observed 13 days before the earthquake with the magnitude of $M_w = 5.6$ at the distance of 140 km from the epicenter. The discovered perturbation may be considered as a complex short-term precursor of the earthquake and the indication of the Earth's crust effect on the surface atmosphere during its preparation.

References

[1] Adushkin V. V., Spivak A. A., Izvestiya. Physics of the Solid Earth **48:3**, 181–198 (2012)
[2] Sobolev G. A., Ponomarev A. V., *Fizika zemletryaseniy i predvestniki [Earthquakes physics and precursors]* (Nauka, Moskva, 2003) 270 p.

[3] Kissin I. G., *Flyuidy v zemnoy kore: Geofizicheskie i tektonicheskie aspekty [Fluids in the earth's crust: Geophysical and tectonic aspects.]* (Nauka, Moskva, 2015) 328 p.
[4] Hoppel W. A., J. Atmos. Terr. Phys. **29 (6)**, 709–721 (1967)
[5] Kulkarni M., Kamra A. K., J. Geophys. Res. **106 (D22)**, 28209–28221 (2001)
[6] Rulenko O. P. Marapulets Yu. V., Kuzmin Yu. D., Doklady Earth Sciences **461:1**, 307–311 (2015)
[7] Rulenko O. P., Volcanology and Seismology **22:4**, 435–451 (2001)
[8] Choudhury A., Guha A., Kumar De B., Roy R., Annals of Geophysics **56 (3)**, 331–340 (2013)
[9] Kachakhidze N., Kachakhidze M., Kereselidze Z., Ramishvili G., Nat. Hazards Earth Syst. Sci. **9**, 1221–1226 (2009)
[10] Spivak A. A., Izvestiya. Physics of the Solid Earth **46:4**, 327–338 (2010)
[11] Yang T. F., Walia V., Chyi L. L. et al, Radiation Measurements **40**, 496–502 (2005)
[12] Guedalia D., Laurent J - L., Fontan J. et al, J. Geophys. Res. **75 (2)**, 357–369 (1970)
[13] Abduvaliev A.K. et al., Doklady Akademii Nauk SSSR **271:1**, 72–75 (1983)

Problem of data interpretation of laser polarization sensing of high-level clouds on the basis of theoretically calculated phase matrices of backscattering by monodisperse ice crystals

Ignatii Samokhvalov[*] *and Ilia Bryukhanov*

National Research Tomsk State University, Tomsk, Russia 634050

Abstract. A technique for determining the microstructure parameters of high-level clouds on the basis of joint analysis of experimental and theoretical data is described. Results of the experiments on atmosphere laser sensing and theoretical calculations of light scattering by spatially oriented ice crystals are shown.

1 Introduction

There is an increasing emphasis on investigations of radiative properties of high-level clouds (HLC) related with their important role in the formation of the Earth's climate and radiation budget. The lack of consideration of accounting the effect of a number of phenomena and processes occurring in these clouds leads to inaccuracies in the calculations of the radiation budget. Spatial orientation of non-spherical ice crystals in clouds is one of such poorly known factors.

The effect of particles in clouds on the transmission of incoming solar radiation and Earth's outgoing thermal radiation determines the need of creating a microphysical model of HLCs considering this parameter. The absence of such models is a consequence of the impossibility of measuring the orientation of aerosol particles in clouds by contact methods because of its violation during sampling, for example using an aircraft.

2 Determination of backscattering phase matrices and experimental data processing

Full information on a cloud microstructure is contained in its backscattering phase matrix (BSPM), which relates polarization characteristics of sensing laser radiation and radiation backscattered by cloud particles. Thus, to evaluate the microstructure parameters of the cloud, it is necessary to determine its BSPM. To investigate special features of cirrus clouds, polarization laser sensing method [1] is used in the high-altitude polarization lidar developer at National Research Tomsk State University (TSU; Tomsk, Russia) [2]. Transmission system of the lidar sends radiation with 4 various polarization states, which

[*] Corresponding author: lidar@mail.tsu.ru

are described by the Stokes vector $\boldsymbol{S}_n$ (n takes values from 1 to 4) into the atmosphere. Receiving system of the lidar implements consequently 4 various instrument vectors $\boldsymbol{G}_m$ (m takes values from 1 to 4) that provides determining the polarization state of backscattered radiation. Unlike existing analogues [for example 3–5], the use of original technical solutions in the lidar developed at TSU allows all 16 elements of a BSPM to be determined experimentally without the use of the properties of pairwise equality of its elements. Wherein, single scattering approximation is used, as it is argued in [6]. Results of the experiments on polarization laser sensing of the atmosphere are compared with meteorological parameters at the altitudes of HLC formation [2]. A schematic diagram of the lidar is shown in Fig. 1.

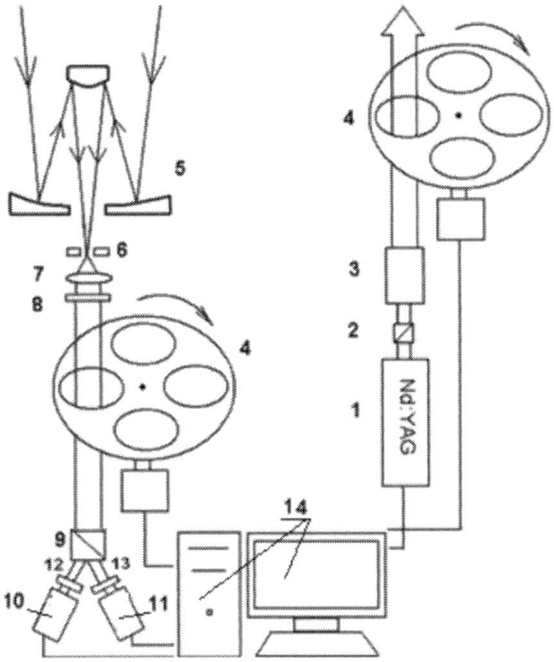

Fig. 1. Schematic diagram of the polarization lidar developed at TSU: *1* is a Nd:YAG laser (λ = 532 nm); *2* is a Glan prism; *3* is a collimator; *4* are units that transform the polarization state of transmitting and receiving radiation; *5* is a Cassegrain mirror objective; *6* is a field stop; *7* is a lens; *8* is an interference filter; *9* is a Wollaston prism; *10* and *11* are photomultiplier tubes; *12* and *13* are electrooptical electrooptical shutters; and *14* is a computer-based equipment for registration and displaying the data [2].

The lidar is built on a monostatic scheme with spaced optical axes of the receiving and transmitting antennas. A Nd:YAG laser with a working wavelength of 532 nm, pulse repetition frequency of 10 Hz, and energy per pulse of up to 400 mJ was used as a source of sensing radiation. The Cassegrain telescope with a diameter of the main mirror of 0.5 m and a focal distance of 5 m served as a receiving objective. A field stop specifying the field-of-view angle of the receiving system, a positive lens transforming a diverging beam into a quasi-parallel one, an interference filter, and a Wollaston prism, at the exit from which two beams were formed with mutually orthogonal polarization states, were successively arranged in the receiving channel along the propagation path of scattered radiation. The beam intensities were registered with Hamamatsu photomultipliers (PMTs); the registration system includes a single-electron pulse counter. Two optoelectrooptical shutters, which suppress backscattering interferences from the lidar near zone, were mounted before the PMTs. System for collection, processing, and displaying the experiemntal data is based on a computer. Lidar is controlled by an operator in semi-automatic mode.

Sensing is performed in the zenith direction. Parallel and orthogonal polarization components of backscattered radiation are measured simultaneously. Both these facts provide the detection of ensembles of aerosol particles with preferred horizontal spatial orientation [7]. One of the special features of such orientation in polarization laser sensing is a low depolarization of sensing radiation upon its interacting with ice crystals oriented horizontally. The presence of a useful signal in one of the polarization channels of the lidar receiving system and its absence in another one during measurements that use combinations of the same polarization elements in the lidar transmitter and receiver allows this situation to be identified. Moreover, a number of special features of such *mirror* clouds are revealed on the basis of the analysis of the experimental data obtained for all operation time of the lidar. Such clouds are formed at the altitudes from 8 to 12 km, have the geometric thickness of not more than 1.5 km, anomalous high scattering ratio $20 < R(h) < 200$, and small optical thickness $\tau < 1$ [2].

Figure 2 shows the examples of the experimental data. Each vertical profile of the lidar signal power is obtained using an accumulation of 500 sensing pulses with a gate length of 75 m. There were two cloud layers: at the altitudes of 7.7–9.5 and 10.6–11.7 km, respectively.

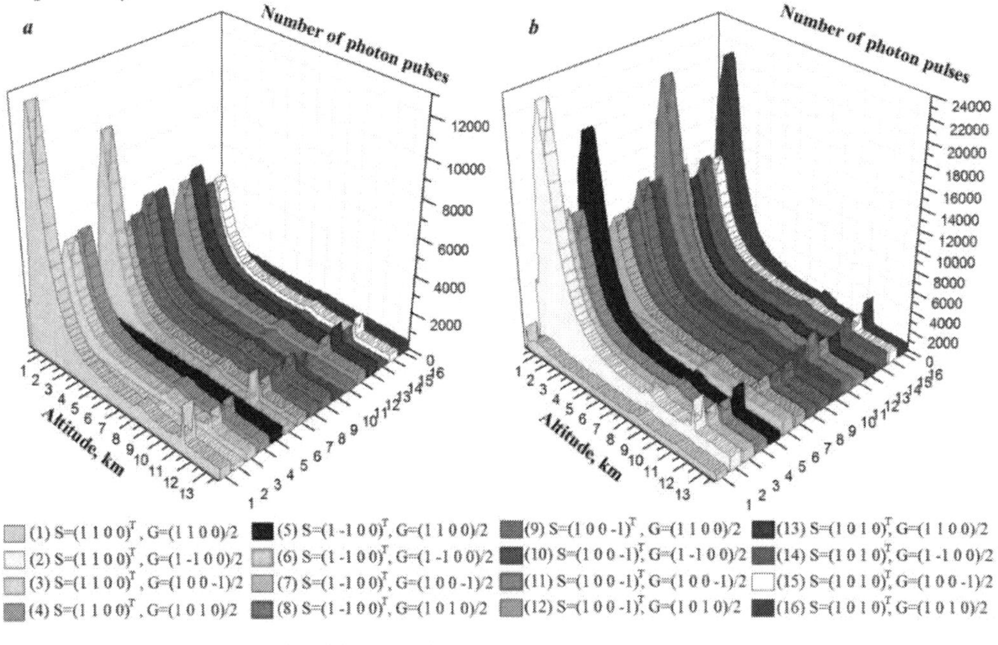

Fig. 2. Altitude profiles of the lidar signal power in the parallel (*a*) and orthogonal (*b*) receiving polarization channels of the lidar (March 24, 2016, 16:50–17:07)

Preprocessing of the experimental data obtained using the lidar developed at TSU includes correction for afterpulsing and erroneous signals of PMTs of the lidar receiving system and reduction of the statistical signal noise level caused by the operation regime of the PMTs in photon counting mode and subtraction of the background noise [8]. Subsequent processing of the experimental data allows determining the altitudes of the lower and upper cloud boundaries, their optical cloud thickness τ, scattering ratio (determined as $R(h) = (\beta_\pi^m + \beta_\pi^a) / \beta_\pi^m$, where h is the altitude, and β_π^m with β_π^a are the molecular and aerosol backscattering coefficients, respectively) [9], and the phase matrix of clouds at the selected altitude [10]. BSPMs determined in the experiments are normalized to the element m_{11}:

$$\mathbf{M}_{\pi}\left(\text{March 24, 2016; 11,25 km}\right)=\begin{pmatrix} 1 & -0,09 & 0,05 & 0,09 \\ -0,09 & 0,64 & 0,53 & -0,29 \\ -0,05 & -0,14 & -0,34 & -0,15 \\ 0,09 & 0,15 & -0,09 & -0,41 \end{pmatrix}.$$

A considerable array of the experimental data is accumulated over the years of operation of the lidar. However, results of processing these data are some data arrays including altitude profiles of lidar signal power, separated geometrical and optical parameters of clouds, as well as meteorological conditions of observations of clouds. At present, a unified base of experimental data combining the listed arrays, systematizing them, and providing quick access to information from each of them is developing. The arrays are extended with new data, when meteorological conditions at the place of the lidar location allow the experiments to be performed.

3 Obtaining BSPMs of polydisperse cirrus calculated theoretically for the comparison with experimental data

Accumulated array of the experimental data obtained using the polarization laser sensing method is statistically significant, however the problem of their interpretation related with the evaluation of the parameters of orientation of crystalline particles is still not solved. This is due to the absence of a strict theory of scattering of electromagnetic waves on an ensemble of spatially oriented ice particles. Our approach to solving the inverse problem – determining the microstructure parameters of clouds – is to compare BSPMs obtained experimentally with ones calculated theoretically. For this purpose, we use the database developed at V. E. Zuev Institute of Atmospheric Optics (IAO SB RAS; Tomsk, Russia) including BSPMs calculated for monodisperse ensembles of ice crystals for various parameters of spatial orientation [11]. The method of physical optics based on the Kirchhoff diffraction theory is used during the calculation of the elements of BSPMs [12].

To obtain BSPMs of polydisperse ensembles of hexagonal ice plates, data from the mentioned database were approximated for further comparison of them with the results of the experiments [13]. Such matrices were calculated as additive sums (with weight coefficients) of BSMPs of monodisperse ensembles taking into account the real parameters of distribution functions of sizes of the cloud particles and parameters of spatial orientation. To approximate the theoretical matrices array, the gamma function was selected [14, 15]:

$$f(a_0)=\frac{\mu^{\mu+1}}{\Gamma(\mu+1)a_{0m}}\left(\frac{a_0}{a_{0m}}\right)^{\mu}\exp\left(-\mu\frac{a_0}{a_{0m}}\right)$$

where $a_0=\sqrt{hD/\pi}$ is the equivalent size of the plates, μ is the variance of the distribution, and a_{0m} is the modal equivalent size. To approximate the orientation angle distribution of the particles, Gaussian function was used [12; 16]:

$$f(\theta)=b\cdot\exp\left[-(\theta-\theta_0)^2/2\theta_s\right],$$

where b is the normalization constant, θ_0 is the modal angle, and θ_s is the variance of the distribution. Since ranges of sizes of the particles are different from those in [14] (Fig. 3), we used another values of the modal sizes.

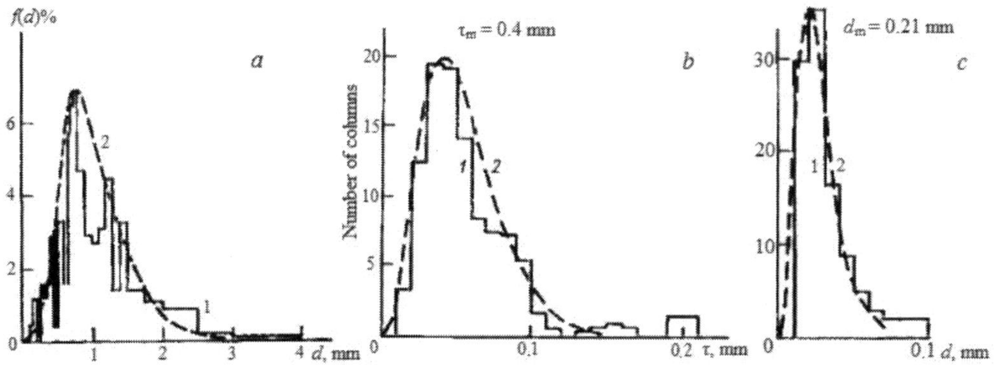

Fig. 3. Size distributions of ice plates (a: d_m = 0.8 mm, μ = 4) and columns (b and c: μ = 3) in cirrus: bar chart *1* is for experiment and curve *2* is for gamma distribution [14]

An extended version of the database developed at IAO SB RAS contains BSPMs calculated for regular shape hexagonal plates and columns [17]. To form the matrices for polydisperse ensembles of ice crystals consisting of the particles of both forms with the parameters of distributions of the particles corresponding to real cirrus, the relation of their fraction from the results of [18] were used. In this work, aggregated data on experimental investigations of microstructure of cirrus obtained in the framework of the projects TWP-ICE (tropics, 2006), ISDAC (Arctic, 2008), and SPARTICUS (middle latitudes, 2010) are shown for ice plates, columns, and crystals of other shapes. The SPARTICUS data [19] were chosen as the closest to the climatic conditions of Western Siberia. Since morphological properties of investigated clouds are not taken into account during processing the data obtained using the lidar developed at TSU, averaged percentages of the crystals were used for obtaining the BSPMs of polydisperse ensembles of aerosol particles (Table 1).

Table 1. Percentages of ice hexagonal columns and plates in cirrus normalized to the total number of regular shape ice crystals [18].

Temperature	$-67 < T < -35$°C	$-40 < T < -15$°C
Columns	47.9 (43.7; 48.7)	45.4 (27.66; 52.6)
Plates	7.9 (30.6; 3.9)	24.3 (57.85; 10.6)

Note: values in parentheses correspond to fractions of crystals in clouds formed from anvils and in another clouds, respectively

4 Conclusions

Functions of distribution of particles in shape, size, and spatial orientation, as well as parameters of these functions corresponding to real high-level clouds were used for obtaining backscattering matrices of HLCs on the basis of matrices calculated theoretically. With this in mind, the described BSPMs can be considered as matrices of HLCs calculated theoretically. Formation of the array of such matrices calculated for various values of parameters of the distributions will allow the inverse problem of estimation of microstructure parameters of real clouds to be solved. Comparison of these data arrays with meteorological conditions at the altitudes of formation of HLCs will allow drawing conclusions about conditions of the formation of clouds with a particular microstructure.

This work was supported in part by the Russian Foundation for Basic Research (grant No. 16–05–00710) and the Ministry of Education and Science of the Russian Federation in the framework of the Tomsk State University Academic D.I. Mendeleev Fund Program (project No. 8.1.12.2015).

References

1. B.V. Kaul, S.N. Volkov, and I.V. Samokhvalov, Opt. Atm. Okeana, **16**, No. 4, 354–361 (2003).
2. I.V. Samokhvalov, B.V. Kaul, S.V. Nasonov, *et al.*, Opt. Atm. Okeana, **25**, No. 5, 403–411 (2012).
3. M.D. Guasta, E. Vallar, O. Riviere, *et al.*, Appl. Opt., **45**, No. 20, 4878–2887 (2006).
4. M. Hayman, S. Spuler, B. Morley, *et al.*, Opt. Exp., **20**, No. 28, 29553–29567 (2012).
5. S.N. Volkov, I.V. Samokvalov, H.D. Cheong, *et al.*, Appl. Opt., **54**, No. 11, 3095–3105 (2015).
6. E.V. Nee and V.V. Bryukhanova, Proc. SPIE, **9680**, 96804M-1–96804M-5 (2015).
7. B.V. Kaul and I.V. Samokhvalov, Atm. Ocean. Opt., **18**, No. 11, 866–870 (2005).
8. S.N. Volkov, *Optical characteristics of high-level clouds from the data on lidar sensing* (Candidate's dissertation in physical-technical sciences, Tomsk, 1986).
9. I.V. Samokhvalov, I.D. Bryukhanov, S.V. Nasonov, *et al.*, Russ. Phys. J., **55**, No. 8, 925–929 (2013).
10. I.V. Samokhvalov, S.N. Volkov, I.D. Bryukhanov, *et al.*, Proc. SPIE, **9680**, 96804G-1–96804G-4 (2015).
11. ftp://ftp.iao.ru/pub/GWDT/Physical_optics.
12. A. Borovoi, A. Konoshonkin, N. Kustova, *et al.*, Opt. Exp., **20**, No. 27, 28222–28233 (2012).
13. I.V. Samokhvalov, S.V. Nasonov, I.D. Bryukhanov, *et al.*, Izv. Vyssh. Ucheb. Zaved. Fiz., **56**, No. 8/3, 281–283 (2013).
14. O.A. Volkovitskii, L.N. Pavlova, and A.G. Petrushin, *Optical properties of crystalline clouds* (Gidrometeoizdat, Leningrad, 1984).
15. V.E. Zuev, *Propagation of laser radiation in the atmosphere* (Radio i svyaz', Moscow, 1981).
16. K. Sato and H. Okamoto, J. Geophys. Res., **111**, D22213 (2006).
17. A.V. Konoshonkin, N.V. Kustova, V.A. Shishko, *et al.*, Opt. Atm. Okeana, **29**, No. 1, 40–50 (2016).
18. J. Um, G.M. McFarquhar, Y.P. Hong, *et al.*, Atmos. Chem. Phys., **15**, 3933–3956 (2015).
19. http://acrf-campaign.arm.gov/sparticus.

 DOI: 10.1051/e3sconf/20161100022

Anomalous changes of vertical geomagnetic field in Kamchatka

Yuriy Moroz[1,*], *Sergey* Smirnov[2], and *Zoya* Nazarova[3]

[1] *Institute of Volcanology and seismology, Far Eastern Branch, Russian Academy of Sciences*
[2] *Institute of Cosmophysical Research and Radio Wave Propagation of the Far Eastern Branch of Russian Academy of Science*
[3] *Kamchatka Branch of the Geophysical Service of the Russian Academy of Sciences*

Abstract. Secular variations of the vertical geomagnetic field at Paratunka (Kamchatka), Kakioka (Honshu), Mamambetsu (Hokkaido) and Patrony (Irkutsk) are considered from 1968 to 2014. Comparative analysis of secular variations showed that from 1968 to 2001, similar variations with the intensity of first hundreds on nT are obvious at four observatories. For the following period from 2001 to 2014, the secular variation at Paratunka observatory differs from other observatories. This disagreement of the secular geomagnetic variation at Paratunka observatory is timed to the increase of seismicity at the depth of 400-700 km in South Kamchatka region. It is suggested that in the result of increase of the seismicity in the region of transition from the upper to lower mantle, physical and chemical processes became more active. That caused formation of a large geo-electrical inhomogeneity which affected the behavior of the vertical component of geomagnetic field.

Investigation of secular variations of the geomagnetic field measured at the observatories in the north-western part of the Pacific ocean, allow us to draw conclusions on the features of geodynamic processes of this region. To do that, we chose the data of geomagnetic field secular variations at the observatories located on mainland (Irkutsk), on Honshu (Kakioka) and Hokkaido (Mamambetsu) islands and on Kamchatka (Paratunka) peninsula (Fig. 1).

There are continuous long-term observations of the geomagnetic field at these observatories (http://intermagnet.org/imos). Synchronous intervals of observations from 1968 to 2014 were selected in the time series. The analysis was carried out for three components of the geomagnetic field (H, D, Z). The main attention was given to Z component which is more sensitive to geo-electric inhomogeneities of a medium. Comparative analysis of vertical component secular variations at Paratunka, Kakioka, Mamambetsu and Irkutsk indicates the following. Similar variations are obvious at four observatories for the period from 1968 to 2001 (Fig. 2). Variation intensity is the first hundreds of nT. We should note that from 2001 to 2014, variations with the intensity of about 100 nT were observed in the annual variations only at Irkutsk, Kakioka and Mamambetsu observatories. This variation is not almost obvious at Paratunka observatory. The question arises, why the secular variation at Paratunka observatory differs from the annual variations of Patrony, Kakioka and Mamambetsu observatories for the latest 14 years.

We refer to the possible nature of secular variations. There are secular variations in the behavior of the magnetic field. They are associated with the change of external ionospheric, magnetospheric,

*e-mail: morozyf@kscnet.ru

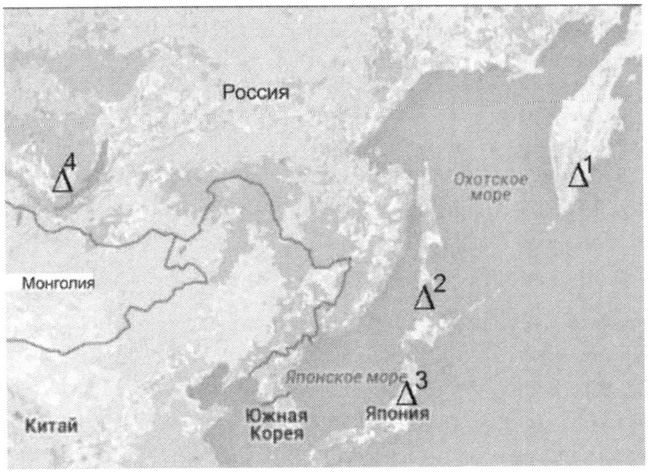

Figure 1. Scheme of location of geomagnetic observatories. 1-Paratunka; 2-Mamambetsu; 3-Kakioka; 4-Patrony (Irkutsk)

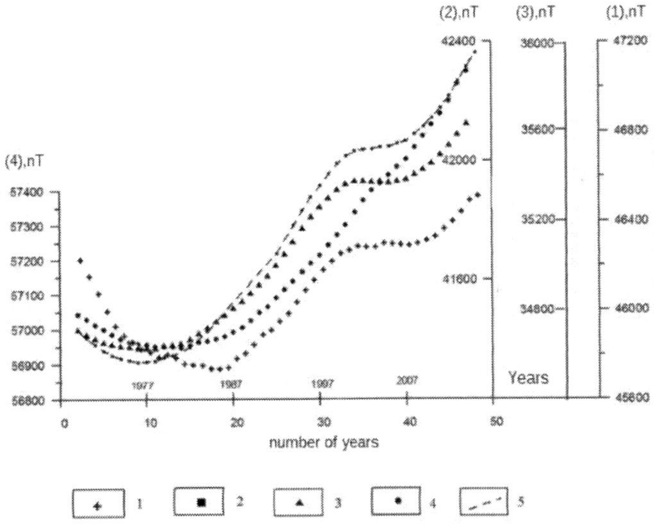

Figure 2. Secular variations of the geomagnetic field vertical component intensity at Paratuka (1), Mamambetsu (2), Kakioka (3), Patrony (4) observatories. 5- absence of values of field intensity at Mamambetsu observatoriy

crust and magnetohydrodynamic sources located in the core liquid part [6, 10]. Changes of the field from crust sources are not large. They are of the first nT and are mainly observed in seismically active regions. Secular variations from the external sources are estimated to be of the first tens of nT. Their intensity is much less than the variations of magnetohydrodynamic sources. It was suggested before that variations with the periods of 11 years and less may exist in depth sources but due to the screening effect of the upper well-conductive mantle do not penetrate to the Earth's surface. Variations with such periods are associated only with external sources. Mechanisms of magnetohydrodynamic generation of field also did not concede generation of variations with the periods less than 10000–100000 years.

However, based on the paper by Karry [11], Braginskiy [3] proved that variations with the periods from 4 to 33 years may be associated with the turbulence pulsations in the core liquid part. According to [11], such variations of nondipole field may be of regional character and occur not over the whole surface of the Earth.

Thus, we may suggest that variations with the intensity of the first hundreds of nT in the secular variations of Paratunka, Patrony, Kakioka and Mamambetsy observatories are regional. As it was mentioned before, the variations are similar during the period from 1968 to 2001 and during the following period the variation almost did not repeat at Paratunka observatory.

We assume that the discovered features may be determined by the change of mantle physical state. Information of the dynamics of rock physical properties at large depths (400–700km) is obtained from the data of long-term seismic monitoring on *www.globalcmt.org/CMTsearch/html* (global) and *data.emsd.iks.ru/dbquaketxt_min* (regional) networks of seismological stations. Seismicity is considered at the distances comparable with the distances to the hypocenters of strong deep earthquakes which could have significant impact on the physical state of the upper mantle. The analysis showed that there were no strong earthquakes ($M \geq 6$) at the indicated depths within the latest 35 years in Japan and the Baikal rift. Strong deep earthquakes ($H \geq 600$ km) occurred only in South Kamchatka region within this period. They were accompanied by multiple aftershocks at the depths of 400–700km (Fig. 3). The strongest one is the earthquake in the Sea of Okhotsk which occurred at the depth of about 630 km with the magnitude of $Mw = 8.3$ (Global GMT). According to the data of the regional catalogue of Kamchatka Branch of Geophysical Service, the energy class of the earthquake is $Ks = 17$ (*data.emsd.iks.ru/dbquaketxt_min*). Its epicenter was located at the distance of about 100 km to the West from South Kamchatka coastline. The earthquake was felt at the distances up to 9500 km [9].

A histogram of earthquakes with $K > 9$ gives the notion on the dynamics of seismicity at the depths of 400–700 km. It is clear from it that increase of seismicity begins in 2001 (Fig. 4). Almost from the same time, the disagreement of secular variations of the vertical geomagnetic field is observed at Paratunka and Patrony, Kakioka, Mamambetsu observatories. As we mentioned before, this disagreement is determined by the fact that the variation in 2001–2014 at Paratunka observatory almost did not manifest in the secular variation.

It is very difficult to discuss the possible nature of the anomalous behavior of the secular variation at Paratunka observatory as there is little information on the exact mechanism of conduction in the mantle. According to the data of seismic tomography, the zone of transition from the upper to the lower mantle in the South Kamchatka region differs by decreased velocity of seismic waves in comparison to the Japanese islands [4, 12]. We can assume the following. In the result of the earthquakes at the depth of 400–700 km, energy released, temperature and pressure changed. It caused the increase of physical and chemical processes in rocks in the zone of transition from the upper to lower mantle. The latest investigations in the area of mineral physics show the possibility of slab depth dehydration in the mantle transition zone and extraction of water-containing liquid [5].

According to [7, 8], the electrical resistivity of rocks at the depths of 400–700 km is 100–5 Ohm·m, respectively. It is assumed that the liquid phase has the electrical resistivity of hundreds and thousands of fractions Ohm · m [2]. Appearance of the liquid phase in the galvanically bound state even in the volume of hundreds of fractions of a percent causes strong increase of rock electrical conductivity [1, 13]. Integrated conductivity of the rock strata at the depths of 400–700 km grows by several orders. Formation of such a conductive depth inhomogeneity in South Kamchatka region is likely to have strongly affected the secular variations of the geomagnetic field. We may suppose that secular variations of the geomagnetic

References

[1] Van'yan L.L., *Electromagnetic sounding* (Nauchniy mir,Moscow,1997) 219
[2] Gordienko V.V. and Loginov I.M., Earth Physics **2**, 35–42 (2011)
[3] Braginskiy S.I., Geomagnetism and Aeronomy **10**, 221–233 (1970)
[4] Jao D., Pirayno F., Liu L.,Geology and Geophysics **51**, 1188–1203 (2010)
[5] Otani E., Chjao D., Geology and Geophysics **50**, 1385–1392 (2009)
[6] Parkinson U., *Introduction into geomagnetism* (Mir,Moscow,1986) 528
[7] Rotanova N.M., Pushkov A.N., *Depth electric conductivity of the Earth* (Nauka,Moscow,1982) 148
[8] Semenov V.Yu., Bulletin of AS USSR, Earth Physics **3**, 60–67 (1989)
[9] Chebrova A.Yu. et al., Volcanology and Seismology **4**, 3–22 (2015)
[10] Yanovskiy B.M., *Earth magnetism* (LGU Pub.,Leningrad,1978) 591
[11] Currie R.G., J.Geophys.Res. **73**, 2779–2786 (1968)
[12] Huang J., Zhao D., I. Phys. Planet. Inter. **73**, 1–15 (2006)
[13] Shankland T.I., Waff H.S., J.Geophys.Res. **82**, 5409–5417 (1977)

DOI: 10.1051/e3sconf/20161100022

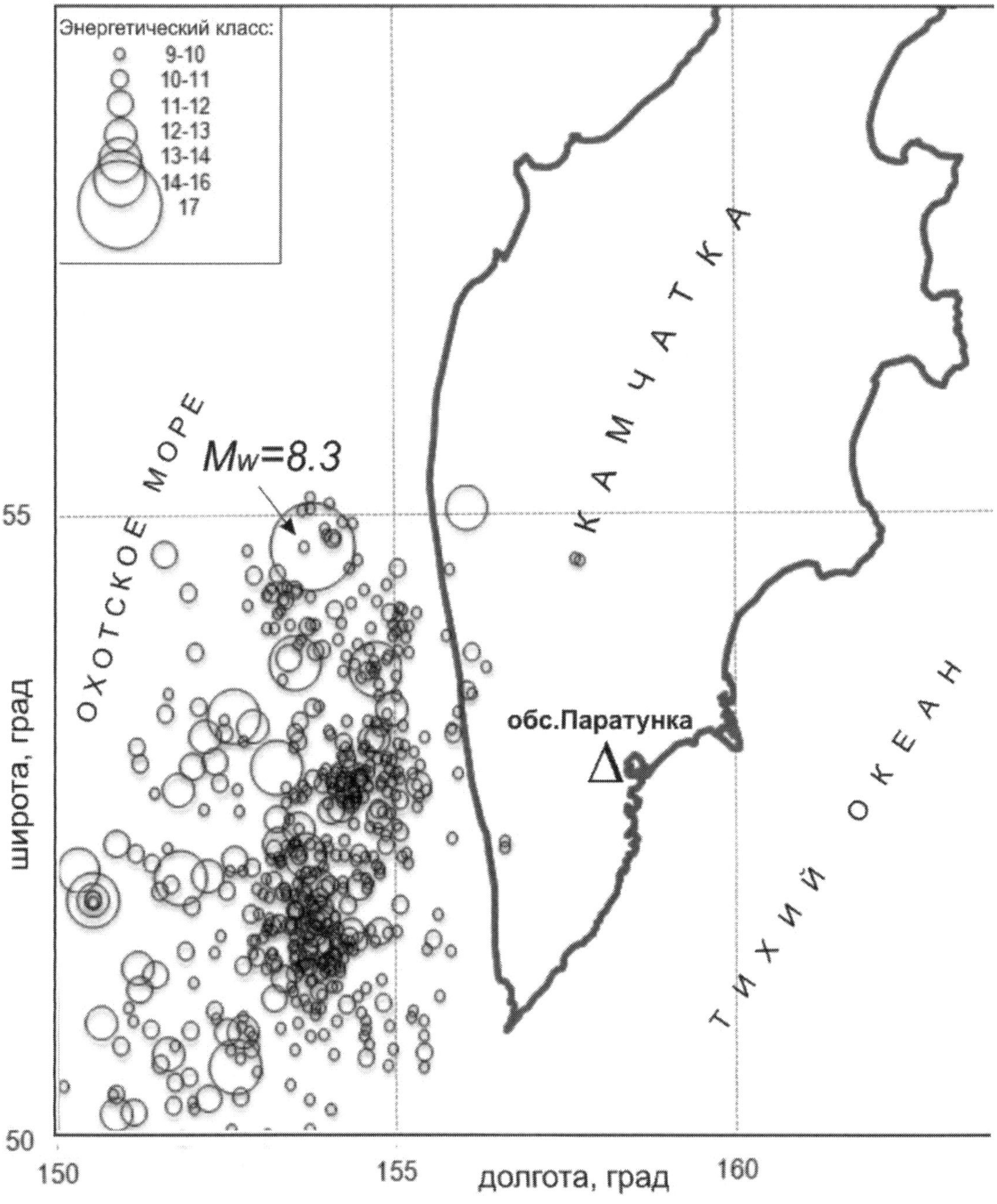

Figure 3. Scheme of location of deep earthquake epicenters (400–700 km) in South Kamchatka region from 1968 to 2015. Triangle shows the location of Paratunka geomagnetic observatory in Kamchatka.

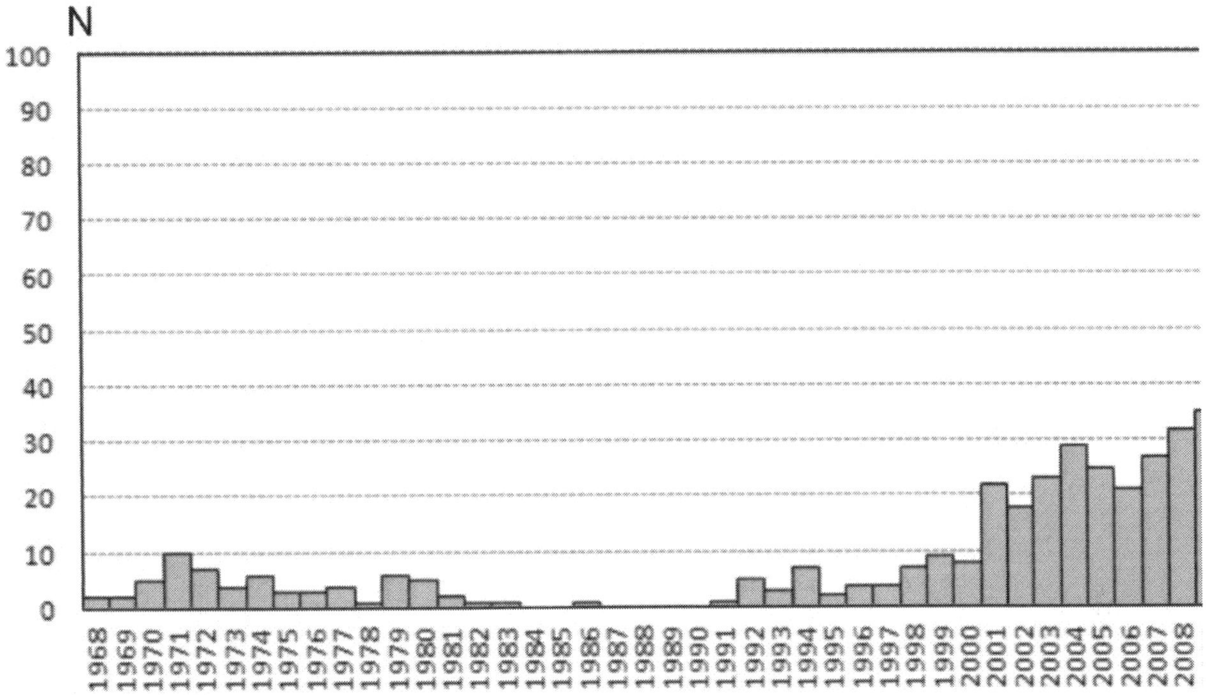

Figure 4. Histogram of deep (400–700 km) earthquakes with the energy class K= 9–17 in South Kamchatka region (Fig.3)

Methods for detection of electromagnetic signals of lithospheric origin

Vladimir Uvarov[1,*]

[1] *Institute of Cosmophysical Researches and Radio Wave Propagation FEB RAS, Paratunka, Kamchatsky Kray, Russia*

Abstract. The paper considers physical conditions of generation and distribution of electromagnetic radiation of lithospheric origin and the peculiarities of registration based on them. Methods for measurements of seismoelectromagnetic activity, parameters and procedures, are suggested.

1 Introduction

The main forming forces of a solid, in particular of a rock, are the forces of electromagnetic interaction. Thus, any change of a form and of a structure of separate pieces is the change of medium electromagnetic structure. It is accompanied both by acoustic and electromagnetic disturbances. In other words, acoustic-electromagnetic emissions of the lithosphere are the manifestations of acoustic and electromagnetic disturbances of the lithosphere during its deformation under the tectonic effect.

2 Instrumentation and methods of investigation

Natural electromagnetic field over the Earth's surface is a mixture of fields of different origin. Its prevailing componnet is associated with atmosperic-lightning activity in the Asian, African and American Centers of the world lightning activity where 100-300 lightning strokes occur every second. The maximum field intensity at the observation site may reach units of v/m exceeding the background vlaue by hundreds of thousands of times [1]. Radiation of lightning origin propagates mainly in the waveguide between the Earth's surface and the ionosphere where the losses are from fractions to the first units of decibels per a kilometer [2]. Thus, when observations are carried out in mid latitudes, the sources of the world lightning centres are located in the far wave zone for SLF-VLF range waves. The component of magnetospheric origin manifests itself the most during magnetic storms. Electromagnetic radiation associated with human industrial activity is significant in densely populated areas and near power lines. At present, natural electromagnetic radiation of SLF-VLF range is applied in global remote sounding carried out by international geophysical projects (TOGA, WWLLN, AWDANET) the results of which are available on their websites. However application of these data in the investigations of lithospheric radiations is problematic.

* e-mail: uvarovvnng@yandex.ru

The lithospheric component is significant in the areas of increased seismic activity during its intensification. The major information on deformation processes in the crust carried over by the low-frequency radiation of SLF-VLF range due to the absorption of high-frequency radiation by a conductive rock. When observing remote lithospheric sources, the propagation path of radiation consists of two parts, they are: from the underground source to the surface and propagation in the atmosphere from the place of radiation output to a detector. Estimates of the cone angle of electromagnetic radiation output for typical lithospheric parameters for this range of waves is within $10^0 \div 20^0$ by a vertical. Thus, the output spot on the survace indicates quite accurately the source projection location. As long as the losses of electromagnetic radiation during its propagation in the Earth's crust are tens of dB/m, signals from lithospheric sources the distance to which is significantly less than the wave length are available for registration. They are located in the near wave zone.

It follows herefrom that in the scales of lithospheric source, the field of lightning origin is homogeneous, whereas the field of lithospheric origin is inhomogeneous. This main difference of statistical characteristics may be taken as the basis of the method for sorting the signals carried by these fields.

3 Sorting of signals

There are several ways to distinguish a LEMR signal from the background noise. Detection is based on such parameters as frequency, amplitude and polarization. Unfortunately, the features for distinguishing of LEMR signals have not been found. Physical restraint (screening) of the size of the area of radiation source and the detector is realized during the measurements in lithosphere cavities, for example in mines or wells. This approach (application of mines) is well known [3] and will not be considered thereby.

Much more effective is the compensation of a noise signal which we consider in some detail. Such an approach may be realized by two ways, scalar and vector approximation of the field.

1. During the scalar approximation of the field (registration of one parameter of the field), the compensation is based on the application of differences of spatial statistic characteristic of the useful and the noise signals.

2. During vector approximation, linear dependence between different components of electromagnetic field is used. Such an approach was applied in magnetotelluric and magnetovariational methods.

Field scalar approximation

The dependence of dipole radiation amplitude on the distance to a source during free space propagation has the following form:

$$A = A^0 \frac{exp(-ifr/a)}{r}\left(\left(\frac{a}{f\,r}\right)^2 + i\frac{a}{fr} - 1\right) sin(\theta) \quad (1)$$

where $a = \frac{c}{2\pi\sqrt{\epsilon\mu}} \approx 5 \cdot 10^9$ m/s —is the constant

c, ϵ, μ —is light velocity, electric and magnetic susceptibility of a medium,

A^0 —is the parameter depending on dipole moment, medium parameters and radiation frequency,

θ —is the angle between radiation propagation direction and dipole moment,

f, r —are the frequency and distance, respectively.

Consequently, the field amplitude in the near zone is proportional to r^{-3} , and in the far zone it is proportional to r^{-1}. Thus, the natural field in a seismically active zone may approximately be considered as a sum of the far (atmospheric-lightning) E_L and near (lithospheric) E_G fields:

$$E \approx E_G + E_L = \frac{E^0{}_L}{r_G} + E_L{}^0 \frac{a^2}{f^2 r_L{}^3} \quad (2)$$

where $E_L{}^0, E^0{}_G$ are the effective amplitudes of lightning and lithospheric sources at the place of generation, r_L, r_G —is the distance from a detector to lightning and lithospheric sources.

To investigate signals effectively, it is necessary for a useful signal to exceed the noises $\frac{E_G}{E_L} > 1$. Unfortunately, in our case this condition is not satisfied. Development of the method for registration and detection of signals from lithospheric sources reduces to the search for such a measurement procedure of field $F(E)$ parameters which would allow us to satisfy the inequality

$$\frac{F(E_G)}{F(E_L)} > 1. \tag{3}$$

In expression (3), the numerator and denominator function arguments in the left part of the inequality have different degrees x . Thus, spatial derivative may be used as such a function. It may be quite easily realized in the form of a measurement process.

We introduce an auxiliary distance in the detector vicinity which satisfies the conditions $x << r_L,\ x \backsim r_G$. Then the relation (3) may be written as follows

$$\frac{d}{dx}(\frac{E^0_G}{r_G + x})/\frac{d}{dx}(\frac{E^0{}_L a^2}{f^2(r_L + x)^3}) > 1 \tag{4}$$

After differentiation of (4) at $x = 0$ it has the following form

$$\frac{E^0{}_G f^2 r_L{}^4}{3a^2 E_L{}^0 r_G{}^2} > 1 \tag{5}$$

Estimates of this inequality in the frequency range of $f \backsim 1 \div 10^3$ Hz at the distances to the center of lightning activity of megameter order of ($r_G \backsim 10^3$ m), air breakdown intensity during a lightning stroke of $E_L{}^0 \backsim 10^6$ V/m, field intensity induced by dislocational motion in a rock of $E_{G0} \backsim 10^{-10} \div 10^{-7}$ v/m [19, 20, 21], show that this inequality is fulfilled more than enough. In other words, in the frequency range of $f \backsim 1 \div 10^3$ Hz a signal from a lithospheric source may be definitely registered and distinguished from strong noise of atmospheric-lightning orignin applying an electromagnetic sensor based on registration of electomagnetic field spatial derivative. We should note that both the signal of lithospheric emission and the signal of induction origin caused by the currents induced by the field of remote sources (near equatorial lightning) may be distinguished by such a way in geological bodies contrast in electrical conductivity. The initial signal causing inductive currents is rejected. The practical realisation of such a sensor is a quadrupole antenna. It is a system of two spaced identical dipole antennas with oppositely oriented dipole moments [4, 5].

Properties of well-sensor

The conditions of wave propagation in the Earth crust get better with depth and vice-versa when approaching to the surface. That is why to investigate the lithosphere wave fields more effectively, it is necessary to receive signals from as great depth as possible. Such approach is possible in mines or in wells covering large territories and diversity of geological conditions.

From the radio physical point of view, a well is 1. a waveguide of longitudinal acoustic and electromagnetic vibrations; 2. a converter of transverse electromagnetic and acoustic vibrations into longitudinal vibrations; 3. a collector of acoustic and electromagnetic longitudinal vibrations of deep layers; 4. a resonance structure. A well casing is a magnetic circuit allowing us to get the magnetic component of electromagnetic emission on the surface. Moreover, it is a magnetoelastic converter of acoustic vibrations into electromagnetic ones (Villary effect). It is a good conductor buried into a poorly conducting environment of a rock. Electric field vertical component generates vertical currents in the pipe material, which in their turn generate a magnetic vortex with an axis along the pipe casing axis.

Data processing for asymmetrical quadrupole

Application of wells in compensative reception by a pair of antennas causes additional difficulties since one of the antennas has a steel casing pipe of the well as a ferrite core which is also a magnetic circuit for channeling of electromagnetic vibration magnetic component to the surface. The second antenna registers only noise signal which is used to detect the useful lithospheric signal. It is difficult to make the amplitude-frequency characteristics of these antennas identical for qualitative compensation of noise signal. Thus, it is necessary to make independent signal registration by each antenna and the compensation should be carried out at the stage of preliminary processing. In order to do that, a weighting function is introduced during the processing. It allows us to compensate the difference of amplitude-frequency characteristics of the antennas. The power spectra of mixture signal and the compensating signal are: $J_0(\omega) =| \tilde{F}(I_0)^2 |, J_k(\omega) =| \tilde{F}(I_k)^2 |$. Here $J, | \tilde{F} |, I_0, I_k$ are power spectrum, Fourier transform module, mixture signal and compensating signal, respectively. As long as the lithospheric signal is much weaker than the noise one, we may find the weighting coefficient in the assumption that the average power spectra of the mixture signal and the compensating signal should be equal with the accuracy to the weighting coefficients: $\widehat{F(\omega)} \cong a(\omega) \cdot (\widehat{I_k(\omega)})$. Here the symbol $\widehat{R}$ denotes the average value of R.

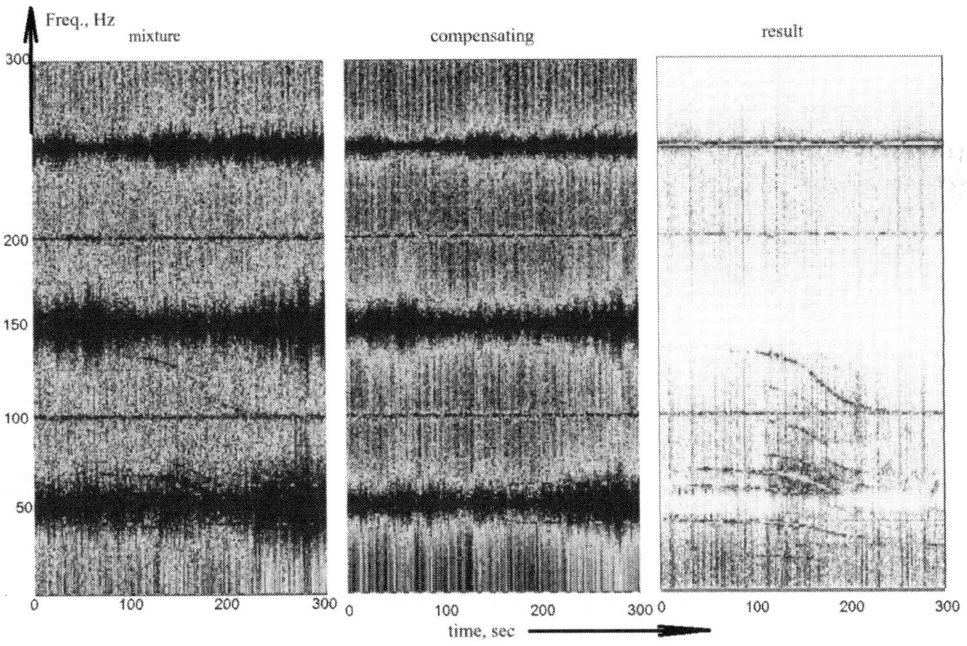

Figure 1. Signal power spectra. A mixture signal is to the left, the compensating signal is in the center, the result is to the right

In order to find the weighting function $a(\omega)$, we apply the least square technique choosing $F(\omega) = (I_0(\omega) - \widehat{a_1 \dot{I}_k}(\omega))^2$ as the efficiency function. Minimizing the efficiency function $\frac{\partial F(\omega)}{\partial a_1(\omega))} = 0$, we obtain weighting coefficients for all the components of power spectrum $a_1(\omega) = \frac{\widehat{(I_0(\omega)\dot{I}_k(\omega))}}{\widehat{I_k(\omega)^2}}$. In this case the power spectrum of the denoised lithospheric signal has the form $I_l(\omega) = I(\omega) - a_1(\omega)I_k(\dot{\omega})$. Fig. 1 illustrates an example of power spectrum of a mixture, compensating signal and the result of detection of a lithospheric signal.

4 Conclusions

Manifestation of seismic events in electromagnetic field discovered at the end of the latest century has become the basis for the development of a new branch of knowledge, seismoelectromagnetism. Electromagnetic phenomena accompanying earthquakes occur in a wide frequency range. The most noticeable manifestations are observed in SLF-VLF range. The complexity of this phenomenon determined by strong effects of the main types of natural radiation, lightning stroke radiation, may significantly or completely be eliminated applying the approaches and methods described in this paper.

References

[1] V. N. Uvarov, Geofizicheskiy zhurnal, **34:6**, 133–146 (2012) (in Russian)

[2] G. S. Golitsyn (eds.), *Geofizika* (Mir, Moscow, 1964), 500 (in Rissian)

[3] G. E. Yakovitskaya, *Metody i tekhnicheskie sredstva diagnostiki kriticheskikh sostoyaniy gornykh porod na osnove elektromagnitnoy emissii*, (Parallel', Novosibirsk, 2008), 315 (in Russian)

[4] V. N. Uvarov, G. I. Druzhin, D. V. Sannikov, Pribory i tekhnika eksperimenta, **26**, 131–137 (2010)

[5] V.N. Uvarov, G.I. Druzhin, V.M. Pukhov, D.V. Sannikov, *Technique for passive location of closely located sources of electromagnetic radiation at the background of powerful radiation of remote sources*, Patent of RF G01V 3/08 No. 2473101, 30.06.2011.

Reversals of the Magnetic Field in one low-dimensional $\alpha\Omega$-dynamo model

Gleb Vodinchar[1,2,⋆], *Anna* Godomskaya[2], and *Olga* Sheremetyeva[1,2]

[1] *Institute of Cosmophysical Researches and Radio Wave Propagation FEB RAS, Paratunka, Kamchatsky Kray, Russia*
[2] *Vitus Bering Kamchatka State University, Petropavlovsk-Kamchatsky, Russia*

Abstract. Described low-dimensional model, that implements the various known regimes of cosmic dynamo systems, in particular, the various regimes of reversals. The source of regular reversals in the model is its internal dynamics. Violation of this cyclicality, failures in the a dynamo, the occurrence regime irregularity reversals caused by chaotic fluctuations in the parameters. This fluctuations are interpreted as the result of spontaneous coherent addition of the higher modes. The simplicity of the model allows you to easily modify it for various celestial bodies.

1 Introduction

For systems of cosmic dynamo (planets, stars, galaxies) are usually considered three mechanisms $\alpha\Omega$, α^2 and $\alpha^2\Omega$, of which the third is the most general [1, 2]. The first two are used when the efficiency of α- and Ω-generator differ sharply. In particular, a strong differential rotation generates a magnetic field described by a $\alpha\Omega$-dynamo.

Known properties of dynamo systems is the presence of reversals - abrupt change in the magnetic field polarity, without significant restructuring of the motion of a conducting medium. Real cosmic dynamo system demonstrate as a regular character reversals (like Sun), and very chaotic (like Earth). The length of the polarity intervals in the geomagnetic field differs by several orders of [3].

The simulation of reversals is intensively developing part of the dynamo theory. Their research is use both direct numerical simulation and simplified models. Direct numerical simulation of the magnetohydrodynamic equations reproduce multiple type of reversals, but does not answer the question of their reason. In addition, the full equations contain a lot of parameters, which estimations differ by several orders, or unknow. Therefore, using simple low-dimensional dynamical systems are trying to explain the physical cause, features, the most important properties of this phenomenon, for example [4–8].

In this paper, we simulate the inversion of the field in a simple model of $\alpha\Omega$-dynamo. This three-mode model with fluctuating of intensities α- and Ω-generators. The source of these fluctuations can be interpreted as a result of synchronization discarded modes of velocity and magnetic field. Such spontaneous formation and destruction of the coherent structures is well known in the theory of turbulence [9].

⋆e-mail: gvodinchar@ikir.ru

2 Model equations

Consider a spherical shell of a viscous conducting liquid in a rotating coordinate system. The origin coincides with the center of the shell, and the Oz – axis of rotation. We will also use the spherical coordinates (r, θ, φ). On the inner r_1 and the outer r_2 boundaries the velocity is zero. This formulation of the boundary conditions is typical for geodynamo and is set for definiteness. For a magnetic field are set vacuum boundary conditions on the r_2.

The physical parameters of the shell is constant. We also believe that the turbulence in the shell is isotropic, and α-effect antisymmetric with respect to the equatorial plane. Therefore we accept scalar parametrization of α-effect in the form $\alpha(r, \theta) = \alpha_0 a(r) \cos\theta$, where $\max|a(r)| \sim 1$ and the coefficient $\alpha_0 > 0$ determines the intensity of α-effect. Now we will consider the problem of kinematic dynamo, but the further we introduce algebraic quenching of α-effect by large-scale magnetic field.

The dimensionless dynamo equation can be written as

$$\begin{aligned} \frac{\partial \mathbf{B}}{\partial t} &= \mathrm{R_m} \nabla \times (\mathbf{v} \times \mathbf{B}) + \mathrm{R}_\alpha \nabla \times (a(r)\cos\theta\, \mathbf{B}) + \Delta \mathbf{B}, \\ \nabla \cdot \mathbf{B} &= 0, \end{aligned} \tag{1}$$

where $\mathrm{R_m}$ - magnetic Reynolds number, and R_α- amplitude of α-effect. Velocity $\mathbf{v}$ is considered a given constant. This type of dimensionless equations corresponds to the choice of radius r_2 as length scale L, magnetic diffusion time L^2/ν_m as the time scale (ν_m - magnetic diffusion coefficient) and some velocity value U and the some magnetic field value B_0.

In the simplest form $\alpha\Omega$-dynamo considered that medium motion is a differential rotation, ie, field $\mathbf{v}$ – toroidal. But we know that in reality this movement is convection and $\mathbf{v}$ includes poloidal component. Moreover, spherically uniform heat flux is excited by convection only poloidal component and toroidal arise only as a result of the Coriolis drift.

Therefore, we set the velocity field as a linear combination of several modes of the free dissipation of the fluid in the shell. All these modes have the form $\mathbf{v}^T_{k,n,m} = \nabla \times \left[R^T_{kn}(r) Y^m_n(\theta, \varphi)\mathbf{r}\right]$ (toroidal) and $\mathbf{v}^P_{k,n,m} = \nabla \times \nabla \times \left[R^P_{kn}(r) Y^m_n(\theta, \varphi)\mathbf{r}\right]$ (poloidal), where index k determines the number of convection layers in the radial direction. The axially symmetric case $m = 0$. Parameters $R^T_{kn}(r)$ and $R^P_{kn}(r)$ depend on the aspect ratio r_1/r_2. We used the terrestrial value 0.35.

Differential rotation correspond to the modes $\mathbf{v}^T_{k,1,0}$. They are consist in the linear capsule of the set $\left\{\mathbf{v}^T_{k_1,1,0}, \mathbf{v}^P_{k_2,2,0}, \mathbf{v}^T_{k_3,3,0}, \mathbf{v}^P_{k_4,4,0}, \dots\right\}$, which is invariant under the Coriolis drift. Because any such mode are driven rest by rotation. Then, in the simplest case, the poloidal velocity component – a $\mathbf{v}^P_+ + 0, 2, 0$. It turns out that the distribution of this mode Coriolis drift direction in the shell volume is well approximated by a combination of four toroidal modes: $\mathbf{v}^T_{0,1,0}$, $\mathbf{v}^T_{1,1,0}$, $\mathbf{v}^T_{0,3,0}$, $\mathbf{v}^T_{1,3,0}$. Therefore, we set the velocity as a combination of these five modes. The coefficients are chosen so that the combination approximates one of the eigenmodes of the Poincare operator . The scheme of calculation of such approximations is described in [10].

For representation of the magnetic field, we will use some of the of Ohmic decay modes $\mathbf{B}^T_{k,n,m}$ and $\mathbf{B}^P_{k,n,m}$. Their structure is similar to the previously described modes of free dissipation of velocity.

We selected of magnetic modes by the scheme proposed in [11].

Let the magnetic field is represented by a linear combination of several modes. Substituting this expansion in the induction equation (1) and apply the Galerkin method. We obtain a system

$$\frac{dg_k}{dt} = \mathrm{Re}_m \sum_i W_{ki} g_i + \mathrm{R}_\alpha \sum_i A_{ki} g_i - \eta_k g_k, \tag{2}$$

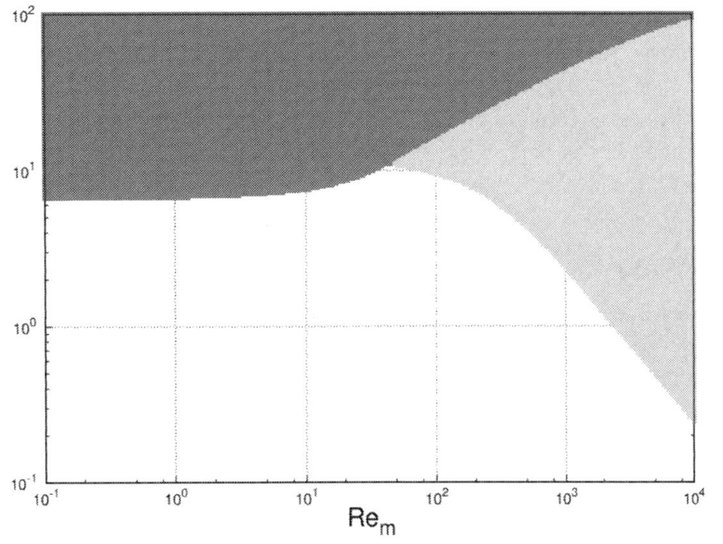

Figure 1. The areas of field generation. Red dots – non-oscillatory dynamo; green dots – oscillatory dynamo.

where $g_k(t)$ – magnitudes of the modes, and η_k – eigenvalues. Matrix W and A are formed by Galerkin coefficients .

Consider the eigenvalues of matrix of the system (2). Let us call the leading and denoted as λ it of them, which has the largest real part. Clearly, the dynamo works if and only if $\Re e\lambda > 0$. Decisions will be oscillating if the number λ imaginary.

In accordance with the approaches of work [11] magnetic field is represented to minimum number of the modes, sufficients to produce oscillatory dynamo.

By varying the parameters Re_m and R_α, we found that oscillating solutions arise, if we use the the following three lower modes: $\mathbf{B}^P_{0,1,0}$ (dipole), $\mathbf{B}^T_{0,2,0}$ and $\mathbf{B}^P_{0,3,0}$. Next, we denote their $\mathbf{B}_1$, $\mathbf{B}_2$, and $\mathbf{B}_3$, respectively.

Of course, the eigenvalues of the system (2) depend on the type of the radial part of the α-effect $a(r)$. We used three ways: $a(r) = 1$, $a(r) = r$, $a(r) = 1/r$.

Note that in the [11] for oscillating solutions authors took 5 magnetic modes for dynamo in the star with a thin convective shell, but they used a toroidal velocity like differential rotation. Most likely, a smaller number of modes in our work can be explained by a more complex structure of the velocity field.

On Fig. 1 shows the areas of oscillatory and non-oscillatory dynamo for $a(r) = 1/r$. Other variants $a(r)$ are given a very similar area, so we continue to limit ourselves this dependence.

Now we introduce in (2) algebraic quenching of α -effect and fuctuations:

$$\frac{dg_k}{dt} = \mathrm{Re}_m(1 + \zeta(t)) \sum_i W_{ki} g_i + \frac{\mathrm{R}_\alpha(1 + \xi(t))}{1 + \sum_j g_j^2} \sum_i A_{ki} g_i - \eta_k g_k, \tag{3}$$

where $\zeta(t)$ and $\xi(t)$ – stochastic processes whit zero means.

As mentioned above, these processes are simulated spontaneously emerging and spontaneously destroying the coherence impact the discarded modes velocity and magnetic field. The structure of the processes is following.

We take on the time axis of the random sequence of points $0 < \tau_1 < \theta_1 < \tau_2 < \theta_2 < \cdots < \tau_k < \theta_k < \ldots$. We assume that the k-th coherent structure is formed at the time τ_k and destroyed at the time θ_k. Then $T_k^{est} = \tau_k - \theta_{k-1}$ - waiting time of formation next structure, and $T_k = \theta_k - \tau_k$ - duration of its existence. The processes $\zeta(t)$ and $\xi(t)$ are zero during the waiting time, and $\zeta(t) = \zeta_k$ and $\xi(t) = \xi_k$ during the time of existence . Here ζ_k and ξ_k independent random variables with zero mean. The laws of the distribution of these variable, as well as T_k^{est} and T_k selected in numerical simulations.

3 Simulation results

Consider the results of computational experiments with the model (3).

From the Fig. 1 we see, that the bifurcation point in the linear approximation (2) is a boundary point of red, green and white areas, i.e. $(\mathrm{Re}_m, \mathrm{R}_\alpha) \approx (49, 11)$.

First, we carried the calculation of the solutions of (3) at $(\mathrm{Re}_m, \mathrm{R}_\alpha)$ with the vicinity of point $(49, 11)$ and zeroes $\zeta(t)$ and $\xi(t)$. In the nonlinear case, the bifurcation point has shifted to $(\mathrm{Re}_m, \mathrm{R}_\alpha) \approx (42.15, 5.45)$. It was also found that the system is highly responsive not only to changes in the parameters, but also to initial conditions.

Therefore, one might expect that the impact of even small fluctuations will switch model between different regimes of dynamo.

In the simulation, we used the exponential law of distribution for the waiting time T_k^{est} and the existence time T_k, and themselves these variables were independent. Mean values $\left\langle T_k^{est} \right\rangle = 5$ and $\langle T_k \rangle = 30$, i.e. time of the existence of coherent structures much less time of their expectations.

The values of jumps ζ_k and ξ_k are uniformly distributed in the interval $[-0.01; 0.01]$ and $[-0.1; 0.1]$, respectively.

The strong differences in the choice of variance of jumps conditioned by the following considerations. α-effect has a turbulent nature, so its response to the restructuring of the turbulence structure i.e. xi_k, can be considerable. Then, disturbance in the magnetic field through the Lorenz force changes the large-scale medium flow. We have described this change jump $zeta_k$. It therefore seems reasonable that the intensity of fluctuations in the turbulence generator should be much higher than in a large scale generator.

By selecting different values of $(\mathrm{Re}_m, \mathrm{R}_\alpha)$ in a small neighborhood of a bifurcation point and different initial conditions, we have received a variety of modes dynamo: quasi-periodic, dynamo-bursts, the disappearance of the field, followed by growth, irregular reversals.

On Fig. 2 shows the examples of the two of realizations. Dimensionless amplitude of the dipole are plotted on the vertical axis.

At the top is mainly quasi-periodic solution. There are only a few failures in the dynamo cycle. This solution is like to the solar dynamo. It is worth noting the almost complete disappearance of the field in the $340 \leq t \leq 370$ and several shorter intervals. There is a certain analogy with the Maunder minimum, the failure of the dynamo.

The solution at the bottom of Fig. 2 is characterized by an extremely irregular reversals, which is typical for Geodynamo.

Thus, in the proposed simple model of dynamo at close values of the parameters can reproduced different modes dynamo, which are observed in real dynamo systems. Of course, all these solutions able to get a lot of writers and earlier as a result of direct number simulation, and the simple models.

The advantage of the proposed model is that it is not associated with the structure of a particular celestial body and does not require knowledge of the distribution of its physical parameters.

The source of regular reversals in the model is its internal dynamics, but violation of this cyclicality, failures in the dynamo, the yield on the regime of chaotic reversals arises due to the restructuring of the turbulence structure.

DOI: 10.1051/e3sconf/20161100024

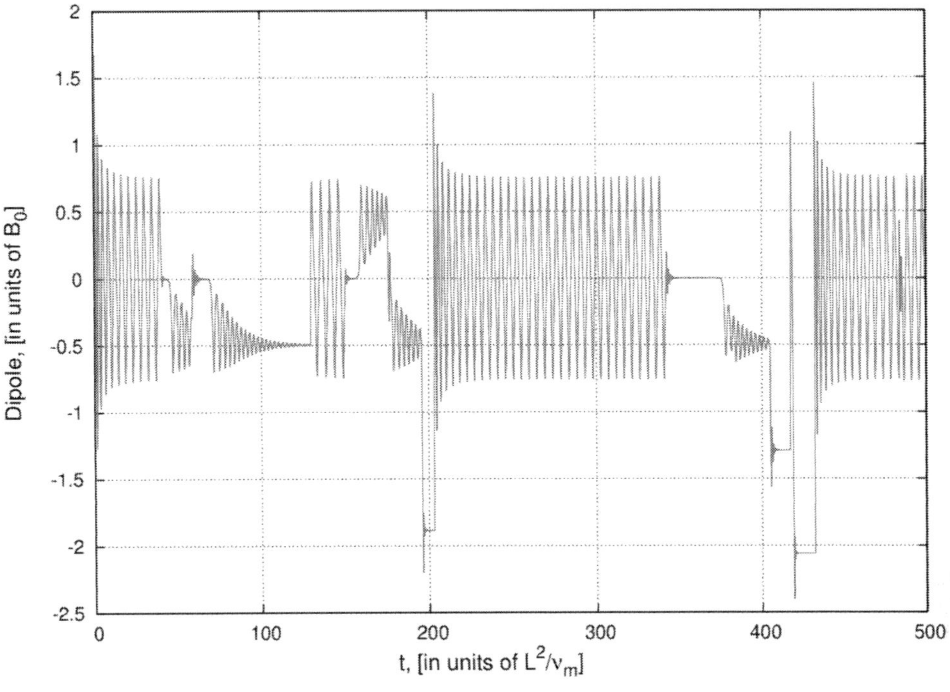

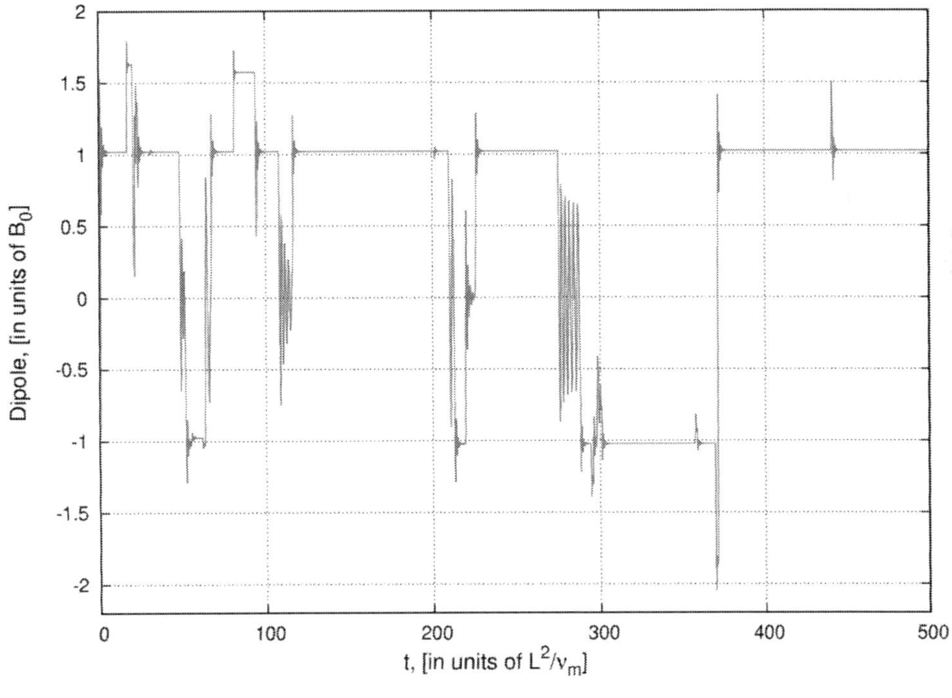

Figure 2. The different regimes of changes the amplitude of the dipole. Top: $(\mathrm{Re}_m, \mathrm{R}_\alpha) = (42.2, 5.5)$. Bottom: $(\mathrm{Re}_m, \mathrm{R}_\alpha) = (42.2, 5.7)$.

4 Conclusion

Proposed in this paper, the model allows to describe various known regimes of cosmic dynamo systems, in particular, the various regimes of reversals.

The simplicity of this model makes it easy to modify it for various celestial bodies. So, depending on the aspect ratio, you can adapt it to the terrestrial planets, giant planets, fully convective stars, stars with a thin convective shell. This will change the parameters of the velocity modes, that may change the set of magnetic modes or their number. Knowledge of the physical parameters of a particular celestial body is not required.

A wide variety of regime may be provided as different distributions of variables T_k^{est} and T_k. For simplicity, we have chosen the exponential distribution for simulation. However, it is known that for turbulence is more characteristic of the power laws. It seems that the introduction of power lows gives an even more complex statistics of reversals, for example, similar paleomagnetic polarity scale.

References

[1] Y. B. Zeldovich, A. A. Rusmaikin, D. D. Sokoloff, *Magnetic fields in astrophysics. The Fluid Mechanics of Astrophysics and Geophysics* (Gordon and Breach, New York, 1983)

[2] F. Krause, K.-H. Rädler, *Mean-filed magnetohydrodynamics and dynamo theory* (Academic-Verlag, Berlin, 1980)

[3] R.T. Merril, M.W. McElhinny, P.L. McFadden, *The Magnetic Field of the Earth: Paleomagnetism, the Core, and the Deep Mantle* (Academic Press, London, 1996), 531

[4] T. Rikitake, *Electromagnetism and the Earth?s interior* (Elsevier, Amsterdam, 1965)

[5] A.A. Ruzmaikin, Comm. Astrophys., **85** , (1981)

[6] D. Schmitt, M. A. J. H. Ossendrijver, P. Hoyng, Phys. Earth Planet. Inter., **125**, 119–124 (2001)

[7] F. Pétrélis, S. Fauve , Dormy E., J.-P. Valet, Phys. Rev. Lett., **102**, 144503 (2009)

[8] L. K. Feschenko, G. M. Vodinchar, Nonlin. Processes Geophys., **22**, 361-369 (2015)

[9] A.V. Kolesnichenko, M.Ya. Marov, *Turbulentnost' i samoorganizatsiya. Problemy modelirovaniya kosmicheskikh i prirodnykh sred* (BINOM, Moscow, 2009), 652 (in Russian)

[10] G.M. Vodinchar, L.K. Feshchenko, Magnetohydrodynamics, **52(1)**, 287–299 (2016)

[11] D.D. Sokoloff, S.N. Nefedov, Numerical Methods and Programming, **8**, 195–204 (2007)

E3S Web of Conferences 11, 00025 (2016) DOI: 10.1051/e3sconf/20161100025
Solar-Terrestrial Relations and Physics of Earthquakes Precursors

Reversals in the six-jet Geodynamo model

Gleb Vodinchar[1,2,★] and *Lyubov* Feshchenko[2]

[1] *Institute of Cosmophysical Researches and Radio Wave Propagation FEB RAS, Paratunka, Kamchatsky Kray, Russia*
[2] *Vitus Bering Kamchatka State University, Petropavlovsk-Kamchatsky, Russia*

Abstract. We describe a large-scale geodynamo model based on hypothesis about 6-cells convection in the Earth's core. This hypothesis suggests indirect data of inhomogeneities in the density of the Earth?s core. The convection pattern is associated with a spherical harmonic Y_4^2 which defines the basic poloidal component of velocity. The model takes into account the feedback effect of the magnetic field on convection. It was ascertained that the model contains stable regimes of field generation with reversals. The velocity of convection and the dipole component of the magnetic field are similar to the observed ones.

1 Introduction

The process of formation of magnetic fields of planets and stars is successfully explained by the theory of hydromagnetic dynamo [1, 2]. The existing models reproduce MHD flows on a small-scale spatial grid at relatively small timescales (~ 50 kyr), or make it possible to calculate only the long-evolution large spatial structures. Clearly, that large-scale spatial structures of convection need to be specified for the models of the second type. This is a key question of the real structure of large-scale convection.

In this paper, we describe a model of geodynamo with six convective cells. The poloidal component of the velocity is determined by the spherical harmonic Y_4^2. In [3], a hypothesis on the six-cells structure of convection was proposed, based on analysis of the splitting of the Earth's free oscillations. It is clear that with such convection two cells are located in the equatorial region, and two cells are located still in each hemisphere. Note that the convective structure of six cells for the liquid core of the Earth has been obtained as a result of direct numerical simulation for some value of the parameters [4]. In this work, three convective cells were located in each hemisphere. The poloidal component of velocity of this convection is associated with the spherical harmonic Y_4^3.

The known properties of real cosmic dynamo-systems is the presence of reversals – change of the sign of the poloidal component of the field. The great interesting are reversals not related to the restructuring of convection.

In this paper, we discuss the question about of the field generation with six-jet Y_4^2-structure, taking into account turbulent α-effect, and reversals in this system.

★e-mail: gvodinchar@ikir.ru

2 Model equations

The equations of geodynamo considering the α-effect in the Boussinesq approximation have the form:

$$\begin{aligned}
&\frac{\mathrm{E}}{\mathrm{Pm}}\left(\frac{\partial}{\partial t} + \mathbf{v}\cdot\nabla\right)\mathbf{v} = \mathrm{E}\nabla^2\mathbf{v} - \nabla p - 2\mathbf{e}_z\times\mathbf{v} + \mathrm{RaPm}T\mathbf{r} + \mathbf{rotB}\times\mathbf{B},\\
&\frac{\partial T}{\partial t} + (\mathbf{v}\cdot\nabla)(T+T_s) = \frac{\mathrm{Pm}}{\mathrm{Pr}}\nabla^2 T,\\
&\frac{\partial \mathbf{B}}{\partial t} = \mathbf{rot}\,(\mathbf{v}\times\mathbf{B}) + \mathrm{R}_\alpha\mathbf{rot}\,(\alpha\mathbf{B}) + \nabla^2\mathbf{B},\\
&\nabla\cdot\mathbf{v} = 0, \quad \nabla\cdot\mathbf{B} = 0.
\end{aligned} \qquad (1)$$

The governing dimensionless parameters of the model are the Ekman number E, the Rayleigh number Ra, the Prandtl number Pr, the magnetic Prandtl number Pm and the magnitude of α-effect R_α. The variable T is the temperature deviation from the equilibrium profile T_s.

It is assumed that turbulence is isotropic and the α-effect is used for scalar parametrization in the form $\alpha = a(r)\cos\theta$, where $\max|a(r)| = 1$. For the velocity, no-slip boundary conditions are applied. For the magnetic field there are specified potential boundary conditions at the CMB and finiteness at the center of the Earth. Finally, the temperature deviation is zero at the ICB and CMB.

For Eqs. (1) we derived of small-mode approximations. The hydrodynamical part of our model is the same as the Lorenz system for convection in a plane layer, but describes six-jet convection structure.

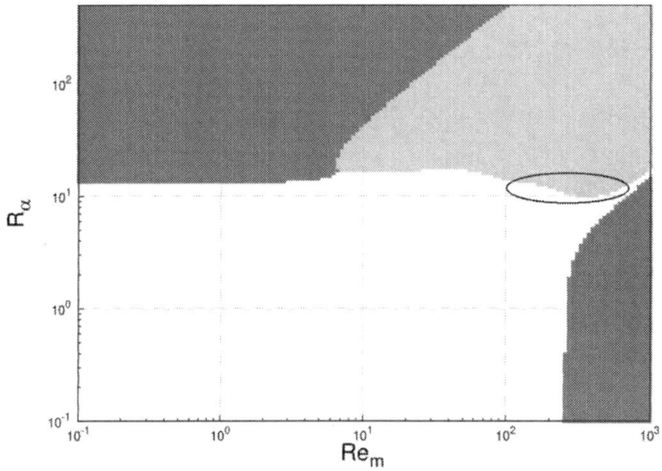

Figure 1. The areas of field generation. Red dots: non-oscillatory dynamo; green dots: oscillatory dynamo.

We use only one velocity mode $\mathbf{v}(\mathbf{r},t) = u(t)\mathbf{v}_0(\mathbf{r})$, where $\mathbf{v}_0(\mathbf{r})$ is a simplest approximation of the Poincaré modes for the spherical shell. The details of construction of this mode are described in the authors paper [5]. For temperature introduce two-mode approximation $T(\mathbf{r},t) = \theta_0(t)T_0(\mathbf{r})+\theta_1(t)T_1(\mathbf{r})$, where $T_0(\mathbf{r})$ matched to the radial component of the velocity mode, and $T_1(\mathbf{r})$ is uniform on the sphere.

For representation of the magnetic field we use some free decay modes, which are selected following scheme borrowed from [6].

First, we order these modes as ascending eigenvalues (dissipation rate). The following sequence was obtained, where the box denotes doublets with identical eigenvalues: $\mathbf{P}_{01}^{-1..1}$, $\boxed{\mathbf{T}_{01}^{-1..1}, \mathbf{P}_{02}^{-2..2}}$, $\boxed{\mathbf{T}_{02}^{-2..2}, \mathbf{P}_{03}^{-3..3}}$, $\mathbf{P}_{11}^{-1..1}$, $\boxed{\mathbf{T}_{03}^{-3..3}, \mathbf{P}_{04}^{-4..4}}$, $\boxed{\mathbf{T}_{11}^{-1..1}, \mathbf{P}_{12}^{-2..2}}$, $\boxed{\mathbf{T}_{04}^{-4..4}, \mathbf{P}_{05}^{-5..5}}$, ...

Next, we selected several modes with the lowest eigenvalues, denoting them by $\mathbf{B}_k(\mathbf{r})$. The magnetic field is written as $\mathbf{B}(\mathbf{r}, t) = \sum_k g_k(t)\mathbf{B}_k(\mathbf{r})$ and substituted into the induction equation (the third equation in Eqs. (1)). The velocity magnitude $u(t)$ temporarily assume constant $u_0 > 0$. Then u_0 can be interpreted as the magnetic Reynolds number Re_m.

Now applying the Galerkin method yields the system

$$\begin{aligned} \frac{dg_k}{dt} &= \mathrm{Re}_m \sum_i W_{ki} g_i + \mathrm{R}_\alpha \sum_i A_{ki} g_i - \eta_i g_i, \\ W_{ki} &= \int_{r_i \le r \le r_o} \mathbf{rot}\,(\mathbf{v}_0 \times \mathbf{B}_i)\,\mathbf{B}_k dV, \\ A_{ki} &= \int_{r_i \le r \le r_o} \mathbf{rot}\,(a(r)\cos\theta \mathbf{B}_i)\,\mathbf{B}_k dV, \end{aligned} \tag{2}$$

where η_i is an eigenvalue of $\mathbf{B}_i$.

In fact, we have obtained a small-mode approximation in kinematic dynamo with six convective cells. Let us denote by λ_i the eigenvalues of the matrix of the system (2). The dynamo works if $\max \Re\mathrm{e}\lambda_i > 0$. The mode with its eigenvalue is the leader, it grows faster than others. It is important that if the eigenvalue of the leading mode is complex, the magnetic field will oscillate.

So, we will gradually increase the number of magnetic modes until an oscillating dynamo appears. Of course, the eigenvalues depend on the parameters Re_m and R_α and on the radial profile $a(r)$ of the α-effect. We used two expressions for the profile, $a(r) = 1$ and $a(r) = r$. It has turned out that the results do not differ qualitatively: the para- meters varied over a logarithmic scale in the ranges $[10^{-1}; 10^3]$. Upon obtaining the oscillating dynamo, we discarded some modes, if this did not qualitatively change the situation and did not increase greatly the generation threshold. As a result, the magnetic modes were the following $\mathbf{P}_{01}^0$, $\mathbf{P}_{03}^{\pm2}$, $\mathbf{P}_{11}^0$, $\mathbf{T}_{04}^{\pm2}$, and $\mathbf{P}_{05}^{\pm2}$.

The areas of oscillatory and non-oscillatory dynamo on the parameter plane $(\mathrm{Re}_m, \mathrm{R}_\alpha)$ are shown in Fig. 1, where $a(r) = r$. Case $a(r) = r$ different small details.

In this kinematic dynamo, an unlimited growth of the field is only possible. To obtain a stable generation of the bounded field, the mechanism of quenching is necessary. In this regard, let us consider the areas in Fig. 1 marked by ellipses. Let us fix $\mathrm{R}_\alpha \sim 15$. Let the point $(\mathrm{Re}_m, \mathrm{R}_\alpha)$ fluctuate along the bigger axes of the ellipse due to the changes of Re_m. It is possible to obtain a stable bounded field generation. Note that in our models Re_m is a fixed value of the magnitude $u(t)$ of the velocity mode $\mathbf{v}_0$. Therefore, we consider a model of magneto-convection with a velocity variable magnitude. To do this, we substitute the considered decompositions for the velocity, temperature and magnetic field in all Eqs. (1) and apply the Galerkin method.

A non-linear dynamic system is obtained, describing the self-consistent six-jet magneto-convection in the Earth's core:

DOI: 10.1051/e3sconf/20161100025

$$
\begin{aligned}
&\frac{\mathrm{E}}{\mathrm{Pm}}\frac{du(t)}{dt} = -\mathrm{E}\mu u(t) + \mathrm{RaPm}S\,\theta_0(t) + \sum_{i,j} L_{ij} g_i(t) g_j(t),\\
&\frac{d\theta_0(t)}{dt} = -Fu(t)\theta_1(t) + Hu(t) - \frac{\mathrm{Pm}}{\mathrm{Pr}}\zeta_0\theta_0(t),\\
&\frac{d\theta_1(t)}{dt} = Fu(t)\theta_0(t) - \frac{\mathrm{Pm}}{\mathrm{Pr}}\zeta_1\theta_1(t),\\
&\frac{dg_k(t)}{dt} = u(t)\sum_i W_{ki} g_i(t) + \mathrm{R}_\alpha \sum_i A_{ki} g_i(t) - \eta_i g_i(t),
\end{aligned}
\tag{3}
$$

where ζ_i is the eigenvalues temperature mode T_i, $\mu > 0$, and $\mu > 0$, S, L_{ij}, F, H are the Galerkin coefficients.

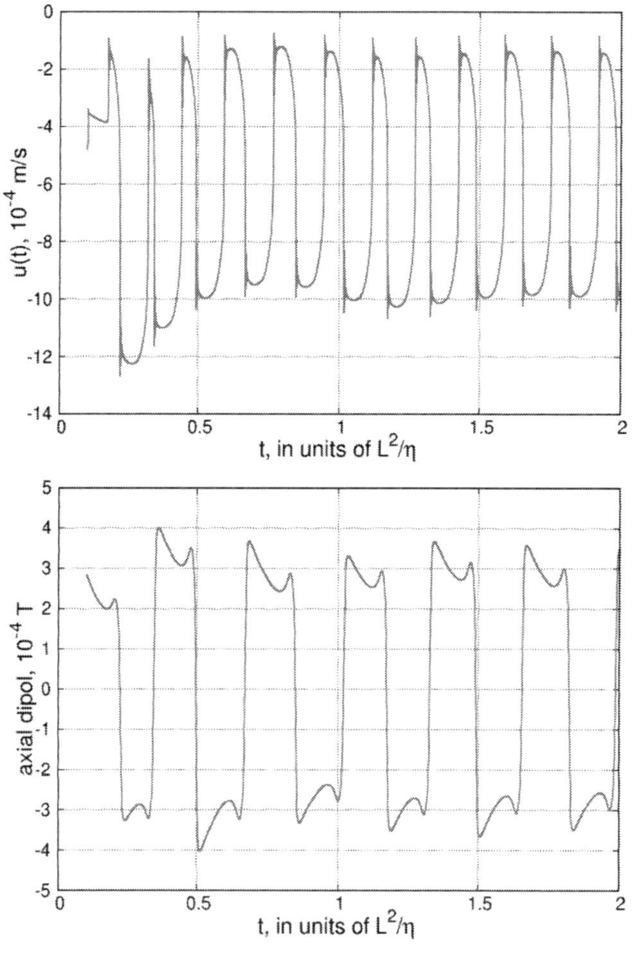

Figure 2. The magnitude of the velocity and axial dipole: Ra = 5000.

3 Simulation results

When performing a numerical simulation with the model Eqs. (3), we used turbulent values of the dissipation coefficients $\nu = 10^2$ m^2/s, $\kappa = 10^{-2}$ m^2/s, $\eta = 20$ m^2/s [7]. The outer radius of the core $r_o = 3480$ km and the angular velocity $\Omega = 7.29 \times 10^{-5}$ rad/s. The relevant parameter values areE $= 10^{-7}$, Pr $= 10^4$, Pm $= 5$ Also, we used $R_\alpha \sim 15$, following the estimations made in [8]. Different Rayleigh number values were used in the simulation, in particular, Ra $= 10^2 \div 10^4$.

We found stable dynamo regimes with reversals at Ra $\sim 10^3$. Fig. 2 shows the graphs for the magnitude of the velocity and dipole part of the magnetic field at Ra = 5000. It is seen that the magnetic field and the velocity are oscillating. However, if the magnetic field undergoes reversals regularly, then the sign of the velocity does not change. Thus, in the model, the reversals have been realized, without convection regime change.

Of course, the numerical results in this simple model should be treated cautiously. However, a characteristic convection velocity value illustrated in Fig. 2 is 5×10^{-4} m/s. This agrees well with the known estimates of the actual velocity of convection $\sim 10^{-4}$ m/s [1]. The characteristic value of the field in the model 5×10^{-4} T is also in good agreement with the extrapolation of the real magnitude of the geomagnetic field at the boundary of the outer core [9]. The polarity interval in the model is 3200 years, which can correlate with mean polarity intervals on the geomagnetic polarity scale [9].

So, the numerical simulations have shown that a stable regime of the magnetic field generation with reversals can be realized with this model. It is important that with these reversals there is no change in regime of convection. The characteristic value of the model convection velocity is in good agreement with the available estimates of the real velocity value in the western drift. The dipole and toroidal components of the model field have the same order of magnitude as the real geomagnetic field.

The reversals of the field in the model have a regular nature in contrast to the real chaotic sequence. Since it is likely impossible to realize chaotic reversals in such simple dynamic model, random perturbations are required for their simulation.

References

[1] C.A. Jones, Phil. Trans. of the R. Soc. of London A: Mathematical, Physical and Engineering Sciences, **358(1768)**, 873–897 (2000)
[2] M. Kono, P.H.Roberts, Rev. of Geophys., **40(4)**, 1013 (2002)
[3] V.V. Kuznetsov, Physics-Uspekhi, **40(9)**, 951–961 (1997)
[4] A.V. Gorelikov, A.V. Ryakhovsky, A.S. Fokin, Computational continuum mechanics, **5(2)**, 184–196 (2012)
[5] G.M. Vodinchar, L.K. Feshchenko, Magnetohydrodynamics, **52(1)**, 287–299 (2016)
[6] D.D. Sokoloff, S.N. Nefedov, Numerical Methods and Programming, **8**, 195–204 (2007)
[7] M.Yu. Reshetnyak, Dr. Sc. thesis, 302 (2003)
[8] A.P. Anufriev, M.Yu. Reshetnyak, D.D. Sokolov, Geomagnetism and Aeronomy, **37**, 141–146 (1997)
[9] R.T. Merril, M.W. McElhinny, P.L. McFadden, *The Magnetic Field of the Earth: Paleomagnetism, the Core, and the Deep Mantle* (Academic Press, London, 1996), 531

AUTHOR INDEX